# FUTURE PRESENCE

## HOW VIRTUAL REALITY IS CHANGING HUMAN CONNECTION, INTIMACY, AND THE LIMITS OF ORDINARY LIFE

## PETER RUBIN

HarperOne
*An Imprint of* HarperCollins*Publishers*

HarperOne

HarperCollins books may be purchased for educational, business, or
sales promotional use. For information, please email the Special Markets
Department at SPsales@harpercollins.com.

FIRST EDITION

Designed by Ad Librum

Library of Congress Cataloging-in-Publication Data is available upon request.

ISBN 978–0–06–256669–0

18  19  20  21  22   LSC   10 9 8 7 6 5 4 3 2 1

*For Kelli, whose presence is everything.*

# CONTENTS

Contents

# INTRODUCTION

# WELCOME TO VIRTUAL REALITY

W HEN THE VIDEO STARTS, all you can see is an older woman sitting in a chair. She happens to be ninety, though signs of her age are obscured by the black box covering her eyes and nose. What you *can* see is her mouth, which hangs open in wonder. When the camera pans away from her, it becomes clear that the black box on her face is connected to a laptop computer, which sits open on the table. On the laptop, you can see two images of a green lawn and trees; that's what the woman is looking at in the black box, which is actually an early version of a virtual-reality headset.

Back to the woman the camera swings, and when she speaks, it's like something out of a Cheech and Chong movie. "Oh, *man*," she says. "It's so *reeeeeal*! Oh, lordy."

"Pretty cool, huh?" the young man holding the camera says.

"It sure is," she says, looking around. Inside of the box, a screen combines the two nearly identical images into a single three-dimensional one that wraps completely around her. If she turns her head to the right or left—or even turns all the way around in her dining-room chair—she can see everything else in the pastoral scene, from a hill in the distance to the stone villa directly behind her. "Is it my eyes," she asks, "or are the leaves blowing? Is this really Tuscany?"

"Someone made this on a computer," another young man in the room says. "This is all made by a computer."

"If I ever explain this to someone else," says the ninety-year-old woman, "they won't believe me."

That sentence, more than any other, gets to the heart of what virtual reality is all about.

Virtual reality (or VR, in the interest of staving off carpal tunnel syndrome) resists description, the way a dream or a memory does. You can tell someone about all the things that you saw and heard while you were wearing a VR headset—how trees looked while you walked in a courtyard on a sunny day, or how waves sounded lapping at the shore—but until someone else has actually tried it, it's just words. It's only once that other person has seen (and heard) those things for themselves that they realize what's possible. Imagine seeing television for the first time, or using a smartphone: sure, it's easy enough to say "it's a box with people inside it," but that gets you only about a quarter of the way there. And forget about explaining Tinder or *Candy Crush Saga* to someone who's never seen a modern phone; they'd just have you hauled away.

New forms of media have always caused seismic shifts in society, from the printing press to the telegraph to a little series of tubes called the internet. But VR isn't simply a new form of media; it sweeps away the barriers of all previous forms. Reading something on paper, hearing a voicemail, and even watching a YouTube video are all enjoyable, yet they're all limited. Each is a representation of the real thing, but it doesn't actually feel like the real thing at all. Television and phones pale in comparison to what VR offers. For centuries, we've experienced art at a remove, imagining scenes in a book or watching movies on a screen. The imaginary worlds within those things are presented to us, and we appreciate them, but we don't actually experience them. Not truly. Now, with the sensory immersion that VR delivers, we have the ability to *become* the art—to be part of a world, even to be a character.

In the five years since that video first appeared on YouTube, VR has grown from that chunky black box into the biggest technological revolution since the smartphone. (And believe it or not, it took about forty-five years for VR to even turn into that chunky black box. We'll get more into that in a bit.) The largest companies in the world have invested billions of dollars in VR—and they did that well before a single device had been officially sold in stores.

Why the optimism? It's not just because VR can create the coolest video games we've ever seen; it promises to upend every industry you can name. Entertainment? Good luck being satisfied watching a movie on a screen after you've had the chance to be *in* the screen, to be there with characters and even have them react to your presence. Travel? No longer do you need to spring for a plane ticket to get some beach time. Education? Take a group of art students to the Louvre without leaving the classroom. VR helps soldiers find relief from post-traumatic stress disorder, and its promise treating chronic pain may one day help it lessen the opioid epidemic. Real estate companies use VR to give clients walkthroughs of homes thousands of miles away. Audi uses it in its showrooms, giving prospective customers the ability to examine anything about any car model, from sitting in the driver's seat to putting their heads inside the engine in order to inspect the inner workings. Through VR documentaries, journalists and governments are putting a face to what would otherwise be abstract in humanitarian concerns—by literally putting those concerns on your face. A couple of years ago, a pediatric surgeon in Miami operated on an infant who wasn't expected to survive; later, he attributed his success with a tricky procedure to the fact that he'd been able to examine a 3-D scan of the baby's heart using VR.

But all that "disruption," as people love to call it, is overlooking the thing that's the most disruptive of them all: the way we relate to each other will never be the same. That's because of something called *presence*.

Presence is what happens when your brain is so fooled by a virtual experience that it triggers your body to respond as though the experience were real. That might mean the fight-or-flight response kicks in when you find yourself in a dark virtual hallway, raising your heart rate and maybe causing a bead of sweat to trickle down your back. It might mean that you feel an upswell of sympathy or compassion when you meet a fictional creature, or that standing in a grand cathedral and hearing a choir gives you chills. It could also mean that when someone—real, or possibly not—moves close to you and whispers in your ear, or simply looks deeply into your eyes, your skin contracts at the base of your hair follicles, leading to a condition scientists call "horripilation" and we call "goose bumps." (For what it's worth, I utterly hate both options.)

Presence is the absolute foundation of virtual reality, and in VR, it's the absolute foundation of connection—connection with yourself, with an idea, with another human, even connection with artificial intelligence. And each one of those things brings with it

4

its own causes and effects. That's what this book is about: looking at what VR can do, how we react in it, and what those reactions mean for how we relate to each other—now, and in the future.

The age of consumer virtual reality finally, truly dawned in 2016. For a year or two before that, people had been able to buy cheap headsets, View-Master-style cardboard and plastic contraptions: you'd drop your phone into one and then strap the thing to your face so that the phone's screen became your window to another world. It was cool, and people called it VR, but it wasn't *quite*. The effect was more like that of one of those old stereoscopes, except it let you spin in a complete circle and see all 360 degrees of a scene. (That didn't stop the *New York Times,* one week in November 2015, from bagging up a million Sunday papers with one of those cheap headsets. Subscribers could use it with any smartphone running the paper's special app and then watch a 360-degree 3-D video documentary about refugee children. Holding the headset to their faces, people found themselves in Syria, Sudan, and Ukraine, seeing the children's lives with a degree of immersion unthinkable a few short years before. Social media filled with images of people using Google Cardboard with awestruck expressions on their faces.)

But in 2016, the first wave of high-powered VR headsets came along that went far beyond 360-degree videos and games. These didn't need a smartphone, but only because they needed a whole lot more power: they plugged in to high-octane desktop computers or gaming consoles. By drawing on the processing power of those computers, as well as some camera-like external sensors, this generation of headsets could induce presence that was fuller and more complete than anything we'd experienced outside of insanely expensive laboratory systems—for the price of a smartphone.

But here's the thing: as impressive as these first headsets are, it won't be long before they seem like clunky relics. The headset that the ninety-year-old woman was wearing in that YouTube video? In VR terms, that super-early version was like an Atari compared with the Xbox One that you can buy today. Wait, no, this one's better: if today's VR headsets are an iPhone, it was like one of those cell phones that connected to a suitcase that you had to lug around, presumably while eating power lunches in suits with enormous shoulder pads. Maybe you were wearing high-top Reeboks! (I'm not judging. At various points in my life, I've worn a glittery Michael Jackson glove, a Malcolm X hat, and jeans I can describe only as "unfortunately large.")

Assuming I actually finish this book without my editor jumping out the window, you'll be able to read it just after the Steven Spielberg movie *Ready Player One* hits theaters. And while we don't have anything quite as immersive as OASIS, the all-encompassing VR universe that everyone uses in the world of the movie (and the Ernest Cline book that it's based on), we're a lot closer than anyone would have thought in 2013, the first time I put a headset on. And these are just the first versions—like iPhones, they'll soon see yearly iterations, and the next decade will bring dizzying leaps forward. Headsets will shrink and lighten, transforming from ergonomic albatrosses into something resembling a pair of sport sunglasses. Screens will leapfrog toward a resolution closer to natural human vision. (High-def TV as we know it features what is known as 1K resolution; high-end 8K TVs will be arriving this year or next; retinal eyesight is closer to 20K.)

And the more people are drawn to VR, the more it will become something as commonplace as an elevator, with uses just as practical and everyday. Which means, of course, people will be using it every day. As millions of people come in contact with VR for the first time, and it slowly becomes a daily part of life—as much as, or perhaps even more than, Facebook and iPhones—it will change the way we live.

No one knows exactly what this might mean for society, but that doesn't mean there aren't doomsday preppers out there. On the massive online community Reddit, an image has been circulating through various VR-related "subreddits," or specific communities, for years: it depicts a young man curled on a mattress in a corner of a room, wearing—like the ninety-year-old-woman—an early version of an Oculus Rift headset. The mattress, like the walls and hardwood floors, is bare. A cable connects his Rift to a nearby laptop, on which you can see what the young man is seeing: a green meadow on a clear day, a rainbow extending overhead from the horizon. "This is where I live now," the title says. (Of the hundreds of comments that have been appended to the image on Imgur, the top-rated one is from more than four years ago: "I bet porn is great with those." Much more on that later.)

Of course the possibility of losing yourself in an artificial paradise exists. It's a classic dystopian scenario, one that science fiction has harped on for decades, in everything from classics like Aldous Huxley's *Brave New World* to prestige TV like HBO's *Westworld*. Every innovation creates a wave of concern about our tuning in and dropping out, and you'll find plenty of hand-wringers today worried that we're barreling toward a nation of slack-jawed VR dropouts like the one in the Reddit photograph.

Much in the way that the internet bred community as well as

escapism, though, VR will have a sociological impact that will dwarf any Chicken Little fantasy of virtual hermits. It'll unfold slowly at first, just like every technological innovation does. But the biggest thing about this technological innovation has nothing to do with technology.

Twenty years ago we may have imagined VR as a way to see and experience a new world—not a way to meet other people (real and artificial) and share transformative, visceral emotions with them. But that's what makes VR, and presence, so seismically important: its ability to facilitate, create, and accelerate intimacy.

Yup. Intimacy.

I get it. The word is a little . . . squishy. It conjures up Dr. Phil moments and the relationship-improving diagnostic tests you find in magazines. ("Go from *50 Shades of Meh* to *Last Tango in Paris!*") Turn the word into an adjective, though, like it was meant to be used, and it goes from a squishy concept to a very real feeling: intimate. It's closeness; it's trust; it's vulnerability and confidence and titillation.

But what has intimacy always depended on? Another person, right? All of this intensity needs someone else who's with you, someone to amplify and reflect the experience, to reciprocate and compound it. These are some heavy sensations, and they're always shared, but now, *for the first time in human history,* we're gaining the ability to induce these sensations without another person—or at least not another real person. Given VR's mind-bending capacity to elicit emotional reactions with a simulation, intimacy can be found with a program or a recording.

Cue the hand-wringers again. Will we all end up marrying robots? Will we forgo our real friendships for our favorite virtual relationships? It's natural to jump to depressing conclusions. But

8

go one step further and we start to see the real possibilities we unlock with access to intimacy—because we're not talking about people shutting each other out. We're talking about people finding something. Maybe it's a feeling of connection that's eluded them for much of their lives; maybe it's therapeutic, cultivating an ability to relate to others in a deeper, more meaningful way. Maybe it's adventure, or arousal, or some other form of fulfillment.

So the real question in all of this is: How will VR affect human interaction? How will it change friendships? Relationships? Marriages? Sex? As VR obliterates the constraints of what we're able to experience, how will it change what we choose to experience—and how we share our lives?

That's what we're going to find out together, me and you.


## WHY ME?


That's a great question. (Also, pro tip for the next time you're watching an interview on television: if someone ever responds to a question with "That's a great question," it's because they hate the question and they're stalling for time. Not that that has anything to do with *this* question. Because yours is legitimately a great one. Have you thought about going into journalism?)

The easiest answer, I guess, is that as a writer and editor at *WIRED*, I've been writing and reading about VR for years. I've been lucky enough to experience things that very few people in the world have, and I've gotten to see next-generation stuff that won't be available to the general public for years. But—and this is important—I'm not a technology writer. Sure, *WIRED* has always been a magazine and website (and social-media account, and Snapchat publisher, and hopefully someday just something you can beam into your brain like Neo learning kung fu in *The Matrix* so you can know every amazing thing about the world we live in) about how technology is changing our lives, but *technology* is only part of that mission: the other part is *our lives*. I've learned enough about VR to explain and describe it to a larger audience, but I'm a culture writer at heart. As much as I'm fascinated by what VR is, I'm much more fascinated by what it can *do*—and what we can do inside it.

Even before I started working on this book, I tried to take as wide a view of the VR world as I could. So yes, I've interviewed the founders of Oculus more times than I can count, but I've also spent time with documentarians and musicians and porn stars and professors and therapists, all of whom are using VR in their own work. And maybe most important, I've spoken with people who don't work in VR at all but have found *in* VR transformative experiences.

All of which is to say, you don't need to know anything about virtual reality in order to read this book. Seriously! The whole point of this book is to help people understand both the technology and its potential. So you're not going to be running into jargon like "asynchronous time warp" and "vergence accommodation conflict" in these pages; honestly, the fact that I actually know what those things are terrifies me. This is a book about showing you, the reader, what the future will look like. What it will feel like. Make no mistake about it: VR won't just change our leisure time. It will change our very culture.

But wait! If you happen to have a passing knowledge of VR, this book is still for you. Hell, even if you're an "enthusiast," if you backed the Oculus Rift's Kickstarter campaign back in 2012 or sold furniture to clear out space for a VR cave at home, even if you go to monthly developer meetups, I promise you that you'll find something in here. You may have already geeked out on the technology, but it's time to explore the human side of the headset. For all VR's mechanical magic, the mental magic it elicits is perhaps more fascinating—and the emotional, cognitive, and psychological reactions we have in virtual worlds promise to change us in some fundamental ways.

So here's how this is going to go. The book, at least as I set out to write it, is a lot of different things. It's a travelogue: a headset's-eye view of the young but still vast world of VR experiences. It's an analysis: thinking about the things I (and other people!) do in VR, and how those things draw on and play with that fragile, complex thing known as the human psyche—and what that enables not just within us, but *between* us. And finally, it's a forecast: starting with what technology allows us to experience now and extrapolating from that to sketch out the changes that we're going to begin to see over the next five, ten, even twenty years of progress.

As for the chapters, think of this book as a slow expansion: starting with the headset itself, each chapter will move a little further out on the spectrum of experiences and relationships. We begin with VR as it exists now—surveying the many offerings for those blushes of intimacy—and then meet the various people who are working to push it forward, as well as the people benefiting from that work. Chapter by chapter, breakthrough by breakthrough, we'll examine a different component of how VR can engender and intensify intimacy. It's a slow mosaiclike composition, through the accretion of people and their stories, of a future no one imagined happening.

## ARE WE THERE YET?

In VR, it helps to know where you're going. Consider this a road map for your journey through the book.

Chapter 1 is a crash course in VR itself: where it came from, how it evolved, and how it works. It traces our collective fascination with the idea of VR, from the books that popularized it to the movies and television shows that made it such an odd cultural footnote of the 1990s. (I'll also get into my own first experience with modern VR.) This chapter dives more deeply into the idea of "presence" and explains exactly what a VR headset needs to be able to do to deliver it. Finally, it takes you through what it's like to use a modern headset for the first time by describing a suite of introductory experiences that come with the Oculus Rift headset. By the time the chapter's done, you won't just know how VR works, you'll know how presence works—and how to find it.

Chapter 2 begins the journey in earnest, by exploring how virtual reality enables us to get in touch with ourselves. We'll get into how the brain deals with stress and anxiety, and how VR can help with that. You'll meet developers who are using VR to make us more mindful, as well as a woman who was one of the first people to explore the promise of VR presence—and is now trying to help other people experience "embodied presence" so that they can experience the changes that she has. Finally, for the first time (but not the last!), I'll take you inside a VR experience that blurs the lines between what's happening inside the headset and what's going on in the real world. Fair warning: there are weird noises. Lots of weird noises.

If you like videos of cute animals, Chapter 3 is for you. (If you hate videos of cute animals, Chapter 3 is still for you, but I might feel a little differently about you.) We're going to start with an Emmy-winning animated VR film that uses presence to induce various emotional responses—except maybe "film" isn't exactly the right word for what people are creating in VR, so we'll talk about that as well. We'll talk about the power of eye contact (and smooth jazz!), and its role in creating intimacy. We're also going to introduce a couple of new ideas: "social presence" and "hand presence," two different facets of presence that unlock entirely new possibilities for interaction in VR and are integral to the rest of the book.

Chapter 4: Okay, we've talked about intimacy already, but here's where we *really* get into it. We'll start with the idea of what intimacy is, and why it's the final frontier of virtual reality. See, up until now, everyone's been fascinated with VR's ability to unlock empathy, but I'll take you through why that word simply doesn't go far enough—and why "moments" matter. You'll also get a short history of VR storytelling and how we're just now unlocking the ability to help you find true presence in someone else's story. And finally, I'll take you to the headquarters of one of VR's most interesting companies and inside an experience that gives us the first hint of how VR can bring us together—even when we're not in the same place.

While my hopelessly nerdy sixth-grade self really doesn't want me to type this, Chapter 5 explores the foundations of "social VR"—starting with me finding the best way to play a game of Dungeons & Dragons. (I know. Believe me, I know.) We'll spend some time in Altspace, one of the first multiuser social VR experiences, in order to discuss what exactly we're all going to *do* together in VR. However, as anyone who's used the internet can attest, bringing people together can go south fast; VR is no different, and in fact, the power of presence might make bad behavior even worse. That means it's time for some not-so-fun stuff: we'll talk about sexual harassment and other toxic behavior, and the outsize effect VR can have on our emotions.

Chapter 6 visits one of the biggest companies in the world, Facebook, to see how it's starting to think about social VR. (Hint: very differently from just about every other social VR company.) We'll take a deeper look at how VR supercharges interpersonal relationships, and how it changes not just what we do together, but how we *remember* it. I don't want to give away too much this early, but it turns out that VR can not only help you learn and retain information, but also create memories that are indistinguishable from real-life memories—which is something we've never had to contend with before.

Now that we've used VR to meet new friends and hang out with old ones, Chapter 7 takes the obvious next step in intimacy and explores VR's romantic possibilities. We'll head to a social VR world we haven't spent time in yet, and I'll introduce you to a couple of people who might just become a couple of people, if you know what I mean. Also, we'll talk a lot more about "avatars," the characters that we create to represent us when we're in social VR spaces. Right now, we're all basically cartoon versions of ourselves, but a lot of people have a lot of ideas about how realistic those avatars should become, so we'll explore why that is and what we can look forward to in the future.

We'll get back to dating, and even more serious relationships, but we're going to take a quick detour for Chapter 8 to introduce a new idea that's going to have a major role in the first few years of VR: "tactile presence." Right now, we don't have a great way to feel stuff in VR—I don't mean emotional feelings, I mean, like, *touching* stuff. (And once more, all together this time: not emotionally touching, *touching* touching.) We'll visit a next-generation entertainment center to see how tactile presence intensifies immersion by marrying the anything-is-possible freedom of VR with the this-is-really-happening undeniability of what your skin registers. Don't worry, though: just because we don't have a great way to feel stuff in VR right now doesn't mean we won't in the future, so we'll take a quick tour of some of the ways our hands and bodies might soon be able to experience presence just as vividly as our eyes and ears and brains do. Then we'll bring it right back to touching each other, by visiting a show that's bringing people together with VR and slow dancing. And zombies. And spaceships. And . . . ah, never mind. You'll read it. Things are starting to heat up.

Every seduction has an outcome—so it's time to talk about VR and sex. (What, you thought we weren't going to get there?) In Chapter 9, I'll take you to the set of a porn shoot to see how the intimacy of presence is revolutionizing everything you thought you knew about the NSFW industry. You'll meet creators and consumers of hardcore VR content to discuss how the technology is upending expectations and maybe even remedying some long-standing problems with the adult industry, and we'll get into how a couple of related industries—camming and sex toys—are harnessing VR for their own ends. If you're in a committed relationship and arching an eyebrow right now, rest assured that we're going to talk about fidelity too. After all, if VR feels real and we remember it as though it's real, does that make simulated sex in VR cheating?

Since "intimacy" is often just a euphemism for sex, it's logical to assume our discussion ends there. But VR isn't all there is.

Given the rise of its cousin *augmented* reality, we'll end the book in Chapter 10 and the book's conclusion by examining how these twin technologies will converge in the very near future—and how mixed reality's ubiquity will affect our lives in the long term. If you love science fiction novels and stoney what-if flights of fancy, this is the chapter for you. Things are gonna get weird.

## A FEW MORE THOUGHTS
## BEFORE WE MOVE ON

As I said at the beginning of this introduction, virtual reality resists description. That doesn't mean I don't try; throughout the book, whenever I discuss a VR experience, I do my best to bring you inside the headset, to give you as complete an account as I can. And for many of the experiences I discuss, you can likely find YouTube videos that people have made recording what they saw in their own headsets; those may be helpful to give you a better sense of a particular piece's visual aesthetic.

But ultimately, as painstaking as they may be, those are both secondhand accounts. Even a recording of VR is only an approximation, appearing as it does on a TV or computer monitor. The only way to really experience VR, to experience presence and the magic it unlocks, is to do it for yourself. So, I urge you: if you're interested, consider buying in. It's cheaper than it's ever been. It will continue to get cheaper, as technology does, but with the imminent arrival of headsets that don't even need a computer, presence is both affordable *and easy* for the first time.

The growth of VR might seem exponential, but what you can see in the outside world—millions of people owning the first generation of powerful headsets, tens of millions more aware of VR—is only the tip of what's happening behind the scenes. The VR industry is moving incredibly fast. New companies seem to emerge every day, many of them purporting to have solved some intractable problem or another, or claiming to have pioneered this or that magical possibility. And as with any rapidly growing technology, that growth is often offset by failure: companies go under, funding dries up. At one point late in the process of writing this book, Altspace, a social VR company that features heavily in Chapter 5, announced it would be closing down. (Cut to me hunched over my laptop in despair, hand hovering over the "delete" button.)

Happily for Altspace's employees and users—and, at the risk of sounding horrible, Chapter 5—the company wound up surviving when Microsoft acquired it, but the announcement highlights something important: as Ponyboy Curtis says in *The Outsiders*, nothing gold can stay. (Yes, he's reading a Robert Frost poem. Which one did *you* read first as a kid?) There's a very real possibility that companies and projects that appear in this book have changed, or may no longer exist, by the time you read it.

It's not just such economic uncertainty that makes the timing of this book tricky. Imagine if someone had written a book about the internet in its early-'90s infancy and explained its potential impact on society. Setting out to write a book about the future of virtual reality is like that: trying to write the story of someone's life when they're still a toddler, wobbling around on chubby little legs. But it's also the most exciting time to document VR's progress and speculate on its possibilities. We're at a crucial point in the technology's development—aware enough of what's going on to witness and analyze this shift as it happens.

Writing about VR has been one of the most surprising, and rewarding, opportunities of my life. As you'll learn in the next chapter, even stumbling into my first hands-on experience with the technology was a matter of right-place, right-time serendipity, and everything since then has felt as charmed as it was welcome. I only hope that I can make you as excited about it—about VR, about presence, and about what they both portend—as I am.

Over the past few years, the technology world has seen a proliferation of "evangelists." Like "ninja" and "rock star," the term means absolutely nothing most of the time, just a buzz-seeking company's attempt to make its marketing executives sound cooler. Well, VR has no shortage of evangelists, but they're not the executives; they're the users. Everyone who has put their ninety-year-old grandmother in a headset is an evangelist. I'm an evangelist too.

You might not be one yet—but I hope this book will help convert you.

# 1

# PRESENCE

## WHAT IT IS, WHERE TO FIND IT,
## HOW TO STAY THERE

N OW THAT WE HAVE introductions out of the way, and judging by the fact that you're reading a book about VR, I'm going to go ahead and assume you're a fantastically curious and sophisticated person. Maybe you subscribe to the *New York Times*—the print edition!—and one Sunday a Google Cardboard showed up packed in with your paper. Maybe you bought a Samsung smartphone and they threw a Gear VR headset in with your purchase, and you took a look at some of the 360-degree videos and games that were bundled along with it. Maybe your son or aunt or friend or someone sitting next to you on a plane insisted that you try their headset. Maybe you were already a die-hard gamer and you'd already priced out and built your own monster of a desktop PC so that you could use it with the HTC Vive you'd preordered. The point is, regardless of when it was, or what exactly you did with it, the idea of virtual reality is at least familiar to you.

But also regardless of when it was or what exactly you did with it the first time you experienced VR, it likely wasn't the first time you'd *heard* of it. We as a society had already had a fling or two with the concept. So while the rest of this book is going to get into where we're going, let's take a second and look back at how we got here.

## A BRIEF HISTORY OF REALITY (THE VIRTUAL KIND)

When Ivan Sutherland was a student at MIT in the early 1960s, he created a computer design program, Sketchpad, that allowed people to use a special pen to draw on a computer screen. That might not seem special in the age of iPads, but in 1963 it was beyond mind-blowing; before Sketchpad, computer graphics simply didn't exist. In fact, the only way to interact with a computer back then was to feed it punch cards like the world's most insatiable parking-garage payment station—so being able to create shapes on a computer screen simply by *drawing* them was tantamount to magic.

But Sketchpad wasn't Sutherland's greatest trick. A few years later, as a professor in Utah, he invented a device called the Sword of Damocles. Scratch that—"device" makes it sound like something you can hold in your hand. This was a complicated and unwieldy set of goggles, suspended from the ceiling like their namesake. In order to use it, you needed to step up to the goggles and allow your head to be tethered to them. (So far, so medieval.) When you looked into the binoculars, you saw two rudimentary computer screens displaying a transparent cube. If you moved your head, the goggles moved along with you—thanks to that handy tether—and the images on the screens changed so that your perspective appeared to change as well. Fifty years ago, the Sword of Damocles became what we now think of as the very first VR headset. It didn't exactly *do* much at first, beyond let you see a cube hovering in space, but Sutherland and a colleague wound up commercializing the technology for use in flight simulators.

It would still be many years before the phrase "virtual reality" would enter the lexicon, however. (At least to describe this type of technology; a French playwright named Antonin Artaud first described theater as "virtual reality" in a 1933 essay.) During the 1970s, Air Force researchers who had been working on flight simulators started to develop flight helmets that could project useful information onto the pilot's field of vision. That project evolved into a program the Air Force called Super Cockpit: a system of helmet, flight suit, and gloves that in the 1980s promised to allow pilots to see—and interact with—3-D simulations of their flight instrumentation and the surrounding landscape, displayed on screens inside the helmet.

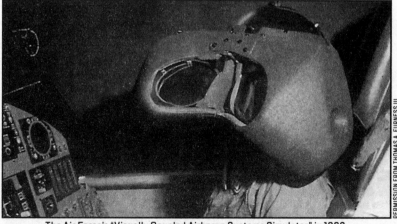

PERMISSION FROM THOMAS A. FURNESS III

The Air Force's "Visually Coupled Airborne Systems Simulator" in 1982.

Meanwhile, scientists at NASA's Ames Research Center in Northern California, a mere two miles from what is now Google's headquarters, looked past our own atmosphere and focused on space. In the summer of 1988, the fruits of that focus appeared on the cover of the space agency's *Tech Briefs* magazine. Above a headline that touted "NASA's Virtual Workstation" was a man wearing a giant white helmet that made him look like a cross between an imperial stormtrooper and a member of Daft Punk. Inside, the accompanying story outlined the promise of the contraption, which NASA called VIVED (Virtual Visual Environment Display), and proposed a future that's by now familiar to anyone who's tried VR: "Imagine having the power to instantly change your environment; to be transported at will to the surface of the moon or a distant star, and yet never physically leave the comfort of your living room. Though it sounds like science fiction, environment-hopping is not only possible but may one day be as commonplace as a drive in the family car." Yet despite the headset explicitly using the word "virtual," the article and the NASA engineer quoted within it referred to the effect only as "artificial reality."

Ground Control to Major Tom: NASA's VIVED system.

In fact, the term we've come to accept was emerging at the same time, a short car ride away. In a small cottage in Palo Alto, tiny VPL Research was developing a pair of goggles, very similar to what NASA was working on, called the EyePhone. (Yes, really.) VPL's cofounder, Jaron Lanier, began calling the technology "virtual reality"—over the protests of his colleagues, who in the Winnebago-mad 1980s thought people would get "VR" confused with "RV." Along with the EyePhone, VPL made a "Data Glove" to control what users saw in the headset, as well as a full-body getup called the Data Suit, which let users see their own limbs in the artificial environment. (The Data Suit was also bright blue and skin-tight, which really pulled the whole Daft Punk thing together.) All that fantasy came at a price, though: the three items, along with the computers necessary to run them, cost more than $350,000.

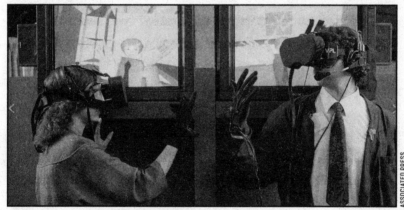

VPL's EyePhones and Data Gloves in action.

VPL would ultimately file for bankruptcy in the 1990s, after the company that funded it foreclosed on loans. But by then both the term and the idea of virtual reality—specifically, a head-worn display that thrust users into an immersive artificial world—had made their way into popular culture. Neal Stephenson's 1992 science fiction novel *Snow Crash* imagined a world in which people used VR goggles to access a series of interconnected worlds called the Metaverse. That same year, through the magic of Hollywood, a Stephen King short story called "The Lawnmower Man" became an amazingly terrible (but not terribly amazing) thriller that updated *Flowers for Algernon*'s simpleton-to-genius plot with a healthy dose of VR. (It also bore almost no resemblance to the original story; in fact, King successfully sued to have his name taken off the movie.)

As the '90s continued, so did the VR touchstones, each one campier than its predecessors. In the cyberpunk thriller *Johnny Mnemonic*, Keanu Reeves used a VR headset and gloves to hack into a Beijing hotel. Nintendo tried to capitalize on the trend with the Virtual Boy, a 3-D video game system so janky and headache-inducing it was discontinued six months after release. And may we never forget the action movie *Demolition Man*, in which a freshly-thawed-after-decades-of-cryogenic-storage Sylvester Stallone offends a woman in 2032 by suggesting that they have sex *not* in VR. (Specifically, he calls it "the wild mamba" and "hunka-chunka." In rapid succession. Four stars.) There were others as well: terrible movie *The Thirteenth Floor*; terrible TV miniseries *Wild Palms*; terrible teen superhero show *VR Troopers*; even a very special (but still terrible) episode of *Murder, She Wrote* in which Angela Lansbury dons a headset to investigate a murder at a VR company. And then . . . it all dried up. VR disappeared, becoming in many people's memory a relic of the 1990s—along with Crystal Pepsi, Beanie Babies, and the Spice Girls.

So what happened? Two things. The first is that all of these visions of VR, however compelling, were just that: visions. We'd seen the photos of researchers wearing their futuristic face computers, and then we'd seen the science fiction movies, and we'd concluded not only that the future was inevitable, but that it was here ahead of schedule, ready to usher us all into its shimmering dreamscapes. Yet, VR in its early forms was awkward and unfamiliar, and when we finally got to experience it from the inside rather than from the outside, what we found was spectacularly underwhelming. VR systems were hugely expensive, were unwieldy and uncomfortable to wear, and were often literally sickening. (We'll get into why that is soon enough.)

But the other thing that happened was that a *different* future arrived—one that was cheaper, more accessible, and cut the legs out from under VR. The year that *Snow Crash* came out, 1992, was the same year the first photo was posted to the internet. By the time the Virtual Boy came out in 1995, an online bookstore calling itself Amazon had opened its online doors, and Microsoft had released the first version of its web browser. Out of nowhere, it seemed, we'd become connected to everything—to information and to each other. Who cared about a weird-looking set of goggles when you could *buy things through your computer*? Even though VR research continued in military and university labs, civilians moved on. It would be almost another twenty years before the promise of VR returned to the public consciousness.

## THE RETURN OF VR

In 2012, I was an editor at *WIRED*, overseeing arts and entertainment coverage. Part of that work meant annual trips to events like Comic-Con International in San Diego and South by Southwest in Austin—and also to E3, a video game trade show held in Los

Angeles every June. My last night at E3 that year, after days of sitting through press conferences and play sessions of upcoming games, I started hearing rumors from colleagues about a mysterious virtual reality demonstration that had been granted to a few lucky people. Once I got back to my hotel I found some pictures of the device, which was being called an Oculus Rift. The thing was clearly a work in progress: silver duct tape stretched across the front of the headset, cables sprouted from three different places, and the whole thing stayed on your head courtesy of a strap ripped from a pair of Oakley ski goggles. But what people had written about their *experiences* with the Oculus Rift, playing an updated version of the classic video game *Doom*, made it sound like someone had actually cracked the code of virtual reality.

Okay, let's fast-forward a year. (This is a book. We can do things like that.) It's now 2013, and I'm back at E3. Since the previous year's trade show, the Oculus Rift has become a hugely successful Kickstarter project, raising more than $2 million from people who wanted their own version of the headset, sans duct tape. I didn't chip in, so I still hadn't tried the Rift, but when a colleague mentioned that he'd scheduled a behind-closed-doors appointment with Oculus, I let him know in no uncertain terms that I'd be tagging along.

This wasn't mere journalistic curiosity. I'd loved *Lawnmower Man* when it came out; hell, I'd shelled out broke-teenager money to see it in the theater. (And I'd shelled out broke-twentysomething money to do the same with the abominable *Thirteenth Floor.*) I'd devoured *Snow Crash* in college, and since arriving at *WIRED* I had similarly feasted on Ernest Cline's sci-fi novel *Ready Player One*, a treasure-hunt adventure that takes place in VR and leaves no '80s pop-culture reference unturned. I may have missed seeing the Oculus Rift the year before, but there was no way I was letting that happen twice.

I still remember that June afternoon in 2013 like it was yesterday. I remember the shirt that I was wearing—a plaid short-sleeved number that my wife still calls "the Marc Maron." I remember walking up the stairs to the upper floor of the Los Angeles Convention Center, to a tiny nondescript meeting room where I met Brendan Iribe, then CEO of Oculus. And when Iribe told us that he'd brought something special to the show, I remember the molded-plastic case that he reached into to fish out the company's first high-definition prototype.

He helped me situate the headset properly, to find that "sweet spot" that would make the 3-D effect as pronounced and comfortable as possible. Inside the headset, everything was dark—then Iribe pressed a button on his computer, and I found myself sitting inside a video game. Like, *inside* the game.

I don't mean inside the game like it was a first-person perspective. I mean that I was in a stone cavern, sitting across from a towering horned creature. If I turned my head to the left and right, I could see the walls that surrounded us. Snow drifted through the air between us; when I looked down, I could see rivulets of lava running along the ground. "Turn around," Iribe said; because I wasn't holding a game controller, I twisted my whole body around. That, as they say, was the aha moment. Because when I turned around in that tiny meeting room, I also turned around inside the cavern, and for the first time I saw how it stretched away behind me. My real-world self—the one sitting in that meeting room—started to grin.

Things moved quickly after that. The following year, Facebook bought Oculus for more than $2 billion. (Fun fact: the sale was announced less than an hour after I had turned in the draft of a cover story I'd written for *WIRED* about Oculus. Cut to me, hunched over my laptop in despair, frantically booking a flight to Southern California for another round of reporting to rework the story.) Other companies jumped into VR, and by the end of 2016 people could get their hands on no fewer than five different solid VR systems—with many more in the works. None of that, though, really explains what VR *is*: what the effect is, how it works, and what exactly you need to do it. So let's take a moment to get into that. Bear with me; I promise you don't need to know anything about "judder," "frame rate," or even "computers."

## THE NUTS AND BOLTS (SIMPLE VERSION)

First, a basic definition. Virtual reality is (1) an artificial environment that's (2) immersive enough (3) to convince you that you're actually inside it. Those numbers aren't there to freak you out, they're just there so we can discuss the ideas one at a time.

1. Really, "artificial environment" could mean just about anything. A photograph is an artificial environment. A video game is an artificial environment. A Pixar movie is an artificial environment. (In some ways, a Pixar movie is a video game already—both are just sequences of computer-generated animations.) Video footage of the room you're sitting in right now could be an artificial environment. The only thing that matters is that it's not where you physically are.

2. An experience doesn't have to look exactly like real life in order to be immersive. (The characters in *Up* were cartoonish,

but remember when Carl ties thousands of balloons to his house to fulfill a promise to his late wife? I'm not crying, *you're* crying.) However, your senses can be manipulated so that you perceive the virtual world to behave the same way you perceive real life. To do that, two illusions need to be maintained: that there's depth in the world, and that you're able to look (and move) anywhere you want to within it, just as you would in life. Creating an illusion of depth is as simple as showing separate, ever-so-slightly different images to your left eye and right eye; your brain does the rest. If you've ever seen a 3-D movie or used a View-Master toy, you've experienced the phenomenon for yourself.

The whole looking-around part is a tiny bit trickier, but it turns out that every single smartphone on the market can manage it. You know how your phone switches from portrait to landscape mode when you're taking a picture—and can stabilize the image even if your hands are shaky? How you can use an app that works like a level to help you hang a poster on the wall? How your phone can double as a step tracker? That's all because of a tiny motion sensor that's embedded in there. And the same kind of motion sensor is embedded in any VR headset in order to know how you're moving your head and then change the image accordingly to reflect your new perspective. (If you're the extra credit type, we're talking about an accelerometer/gyroscope combo that's commonly known as an "inertial measurement unit." They cost fewer than five bucks, and they're about a tenth of an inch on each side. Science, man, science.)

So if in VR you're standing in the doorway of a house, you should be able to look down—like, actually look down with your head in real life—and see the carpet under your feet, or to the right to see the posters on the wall, or even behind you to see the lawn and the cars parked on the street.

Why does this matter so much? Because we've never been able to do it. When you're watching a movie, the thing you see changes only because the director changes the camera's perspective. When you're playing a video game with a first-person viewpoint, you use a game controller to change the camera's perspective. (You may have more agency than with a movie, but a game controller isn't your face; it's an interface.) Humans tell each other stories, and every story we've ever depicted visually (a painting, a television show, a play) has put that story inside a frame (a canvas, a screen, a stage's proscenium). With VR, your perspective is now at the center of an artificial world; virtual reality takes you *through* that frame and into the world itself. You're inside the movie, the game, the painting, the story.

3. Let's put all this in perspective. No matter how good a VR headset is, you're not going to forget that you're *not* in real life. That's because all screens, from the old tube TV you had as a kid to your smartphone to a super-high-def monitor, are made up of tiny rectangles called pixels. The closer those screens are to your eye, and the less densely those pixels are clustered, the more obvious the pixels are—and the more obvious it is that you're looking at a screen. On a VR headset, the screen is less than two inches from your face, which means that in order to give you an image that's indistinguishable from real-life vision, it needs to display more than two thousand pixels *per inch*. That's about ten times more densely packed than a MacBook's "retina display"—and it's also going to be a number of years until those kinds of displays are easily available. (However, I've recently seen prototypes of a technology that uses tiny microdisplays packing in nearly three thousand pixels per inch. A small sensor inside the headset watches where

your eyes are looking and then projects that microdisplay exactly where your pupils are focused. It's not in a commercially available headset yet, but the effect is jaw-dropping.)

But here's the thing: that doesn't matter. Yes, your rational brain might know that you're using VR, but your rational brain can also be tricked—or at least overruled. That phenomenon is the cognitive con known as "presence," and it's the key to just about everything you're going to read in this book.

## PRESENCE: THE "IS" OF VR

When VR is working well, your physical senses tell your brain that you're really experiencing the thing you're virtually experiencing, and your brain prompts your body to respond in kind. That's presence. The International Society for Presence Research (fancy, right?) describes it using a nearly three-thousand-word manifesto (crazy, right?), but this tiny bit of it is really the key: "the individual can indicate correctly that s/he is using the technology, but at *some level* and to *some degree*, her/his perceptions overlook that knowledge . . . as if the technology was not involved in the experience."

That's a pretty antiseptic way of phrasing it; what actually happens is much more visceral than that. As a very simple example, let's say that we're standing together in your living room. (And I must say, your taste is superb. Love that coffee table.) I give you a VR headset, and when you put it on, you find yourself standing on the ledge of a skyscraper. When you look down—careful!— you can see that the floor is hundreds of feet below you; there's

31

another building across the street, but it's probably a hundred feet away itself. Something like this.

Go ahead, take a step.

Then you hear me—don't forget, in the real world I'm still in your living room with you—say, "Okay, step off the ledge." You know you're in VR, and you know that if you actually lift your leg and step forward, you'll just put your foot down on your living room carpet. But that's your rational brain talking, the part that's able to reconcile the idea that you're seeing one world and standing in another. There's also your reptilian brain. And your reptilian brain has decided that you're not actually standing in your living room but on the ledge of a giant skyscraper.

Why shouldn't it? That's what your eyes are seeing. The skyline has depth just like a real room, and when you turn your head you can see that it surrounds you just like the real world would. The fact that it looks like a video game and not *exactly* like real life doesn't matter; what matters is that the part of your brain that cares only about survival sees the drop to the street below and says, "OH *HELL* NO YOU'RE NOT STEPPING OFF THAT LEDGE!"

Depending on how afraid you are of heights, your sympathetic nervous system might kick in, causing your heart rate to rise and your palms to sweat. Either way, you will find it *very* difficult to take that step forward. In your brain, you're there on that ledge; that's the very core of presence.

While it sounds easy enough, presence is actually pretty tough to come by. In fact, the two bestselling VR headsets up to this point can't deliver it. Google Cardboard and Samsung Gear VR use your smartphone to provide the screen and motion tracking. (The headsets themselves are basically empty shells with a pair of magnifying lenses to make sure your two eyes are looking at the separate images.) That's great for your wallet, but it comes with a limitation: your phone can follow only your head's orientation, not your body's position.

Why is that a problem? Well, for real presence, you need to be able to do more than just move your head around—you need to be able to *move your whole body*. Imagine your head at the center of the diagram in the figure.

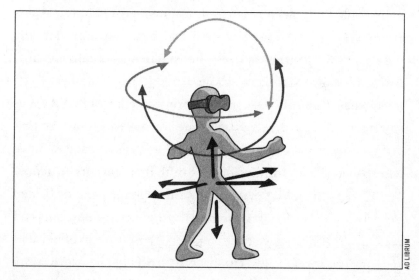

AD LIBRUM

Those three curved lines are the type of motion that any cheap sensor can track: looking up and down, turning your head to the left and right, and tilting your head so that your ears drop toward your shoulders. But it can't track the straight lines, the actual locomotion. Pretend we're standing in your living room again, and this time you're wearing a mobile VR headset. In the headset, you're standing in front of a desk, and there's a piece of paper on the desk that you can't quite make out. You want to be able to read the piece of paper, so you lean over—but when you do, nothing happens. The desk is in the exact same place, despite the fact that in the real world your head has moved down and forward.

At the very least, that's annoying. The virtual room is no longer behaving like a real room might, and so your experience is much less immersive. However, this also introduces a new problem. When you lean over, the balance system in your inner ear senses it and expects your perspective to change as well—but when it doesn't, that conflict between your eyes and your ears sets off an internal alarm. *Maybe you've been poisoned,* your brain tells your body. Your body, which generally listens to your brain, reacts appropriately. Maybe you feel a tiny bit dizzy, maybe even mildly queasy. It's like motion sickness, but you're not actually moving. In VR, it's called simulator sickness, and it has a number of different causes. In this case, it's easily prevented: mobile VR experiences are designed to be used while you're sitting down so that you can rotate your head freely but not feel encouraged to move it in space. Still, it's the kind of constraint that prevents presence.

Unlike using a VR device on your mobile phone, a dedicated PC-driven headset has its own embedded screens and sensors and connects to a high-powered computer. It also uses external sensors that track your headset in space, which lets you move inside of

34

a VR environment and thus makes presence much easier to come by. In order to really illustrate what presence is, and the reactions it can elicit, let's look at what was many people's first experience with it: a VR orientation that launches when you first start up the Oculus Rift, which became the first high-end VR system available when it went on sale in 2016.

## ENTER THE DREAMDECK

Again, picture yourself in your living room, standing in an expanse of whiteness on a translucent circular platform. Well, not *standing*; you might be standing in real life, but none of your body made it into VR along with you. Here, you're just a pair of eyes. "Welcome to Oculus Dreamdeck," a woman's voice intones. Dreamdeck is a suite of short vignettes designed to be a first-timer's tour of VR. As you might have guessed from the whole no-body thing, you're a passive observer here. The whiteness fades to black, and a new scene fades in.

You're in what looks like the interior of a submarine. Ahead of you, you see a periscope; the floor is tiled. It's dim, and the few fluorescent light panels in the ceilings and walls don't do much to illuminate the cabin. You look to your left and see a console of lights and instruments, their words clearly legible: TRIM PANEL, HULL, RIG FOR DIVE. Behind you, a low archway leads out of the cabin. You take a step toward the periscope and lean in so you can examine the details better. After about thirty seconds, the scene fades out and is replaced with another.

There's no room here that you can see, just blackness. A dinosaur stands in front of you, lit from above. It's in profile, and despite unmistakably being a *T. rex*, it's not much taller than you. Something about it feels like an exhibit, so you're bolder than you might be otherwise; when you approach it, it swings its head toward you and softly roars, and its tail sways back and forth. You're so close you can see the pebble-like texture of its face scales, the color of its tongue. It's not a threat, but a curiosity—then it, too, fades out.

While the first two scenes were nearly photorealistic, with textures and graphics that equal those of any big-budget blockbuster video game, now you're somewhere markedly different. It's a bucolic meadow in a valley, with a stream passing through it, but—well, remember tangram, the childhood craft game that challenged you to make animals out of flat cardboard shapes like rectangles and parallelograms? Yeah, you're basically in Tangram World. The trees look like 3-D versions of a child's drawing: green triangles on top of brown rectangles. A fox, a moose, and a rabbit are gathered around a campfire with you—and they're all blocky approximations of themselves, from the fox's white-tipped tail to the rabbit's cutely drooping left ear to the triangular embers that float upward from the fire. But despite the animals' lack of eyes and the rudimentary graphics, there's a feeling of serenity. The fire is crackling, the brook is babbling, the blue sky overhead is dotted with puffy clouds (or at least tangram-angular approximations of puffy clouds). All is well. In fact, when you look at the animals, it seems like they look back at you—the moose seems almost to nod in greeting. They're here with you, you'd swear it. Just as you look up to notice a cardinal in a nearby tree, the scene fades out. *What's next?* you wonder, only to find yourself . . .

. . . on the ledge of a skyscraper at night, peering down to see the street hundreds of feet below. The tangrams are gone, and everything is shockingly realistic—if hewing to a distinctly art deco aesthetic. Looking around, you realize you're in the middle of an urban skyline, with similarly tall buildings all around you. A dirigible hovers overhead to the left; to your right, you see the span of an enormous bridge. Before you have a chance to soak in all the details surrounding you, though, the scene goes dark.

When a new environment fades in, you're standing in front of a large, gold-framed mirror. In the reflection, you can see a white marble table behind you set for tea; green and yellow lanterns hanging from the ceiling give the room a warm glow. The face in the mirror looks like an ancient statue of an old man, with a long beard. You turn your head this way and that, and your reflection moves perfectly along with you. You lean toward the mirror to get a closer look . . . and with a burst of light, the face transforms into an ornate mask, festooned with gold embellishments and empty eyeholes, a coy half smile frozen in place. Still, though, the mask moves as you do, and even though it's incapable of changing its expression, you're convinced that it's you. But just as quickly as the mask transformed, it does so again, and then again and yet again: a small purple safe, its gold embellishments giving it the look of a smiling face; a carving of a sun; a red balloon with a woman's visage drawn on it. None of the objects look remotely like you, but they mimic your movements so accurately that it's impossible not to see yourself in them.

This time, after the scene fades and a new one appears, you're not even on Earth anymore, but on a snow-covered planetscape against an alien sky. The creature in front of you is short and gray, its heavy-lidded eyes and comically downturned mouth making it look like the grumpiest stoner on Zarblox 4. When it notices you, though, its eyes go wide; its iridescent brow furrows as it examines you. You step to the side, and it turns to watch you as you crouch down and examine the planet's rocky surface. The creature says something, so you stand up and take a step closer, then closer still. How can you not? It's the first alien you've come face-to-face with! Thankfully, it doesn't seem to resent the intrusion, though it does seem to *tsk* a bit with disappointment at your inability to speak its language.

The next scene is just as alien, but in a way you might recognize from watching science shows as a kid: you've been shrunken down to the microbial level, the world the same monochromatic gray as in an electron microscope readout. A mite towers over you; huge red blood cells and bacteria float by. This scene goes by fast, with just enough time for you to get your bearings and look around.

Now you're looming over a tiny island city, able to lean in and examine its many buildings and goings-on. Everything here seems to be made out of paper; there's depth, but it's cartoonish and utterly charming. There's the air-control tower and landing strip; there's a snowcapped mountain and the city's tallest skyscraper; there's a helicopter hovering in the sky. You notice a small office building on fire, so you lower your face to its level—and see a fire truck with its ladder extended, a tiny paperlike figure in its bucket spraying water at the source of the fire. In the foreground, cars drive by a residential neighborhood, and another paper resident reclines in a lawn chair while his charcoal grill sputters. It's like the world's most immersive model landscape, a Richard Scarry book come to life; you could happily spend an hour in here peeking into windows and watching city life unfold.

You don't have an hour, though, and it's time for another scene. You're back in a featureless white environment, watching two giant robotic arms examine a rubber-duck bath toy, conduct an invisible orchestra, and ultimately engage in a harmless lightsaber fight using conductors' batons. It's as silly and engaging as any Pixar short, with the machines taking on undeniable personalities—and it's nearly as long, this one clocking in at a minute and a half. There's one crucial difference, though: you're not in the front row of a theater, but in the middle of the action. The scene fades for the finale.

Now you're in a long hallway in a museum. Against one wall a dinosaur fossil is embedded in sandstone; against the other a *T. rex* skull sits on a platform, a poster above it declaring REX LIVES. Outside the windows, it's night. Suddenly, you see something around the corner at the end of the hall. It's the *T. rex* from earlier in the Dreamdeck—except either you've shrunk or it's hit a growth spurt, because as it lumbers toward you, it's clear that you're no match for this thing. It stops in front of you, then lowers its head and lets out a mighty roar, vapor and spit flying from its mouth as it does. (This is one of those times to be happy that smell isn't in VR yet.) It turns one yellow eye to examine you and then continues on its way, moving directly over you as it does. Its underbelly passes over your head, its rear legs striding by on either side of you, its tail nearly brushing your head as it recedes behind you into the darkness.

And with that, Dreamdeck finishes. You've been underwater and in the air. You've been examined by aliens and dwarfed by insects. You've befriended kindly animals and cowered before other, bloodthirstier animals. You've made small, wordless, but very real connections with other creatures; you've believed that a floating red balloon was actually *you*. You've been charmed, terrified, intrigued, and disoriented. You've spent what feels like an hour getting your first taste of presence. In reality, though? It's been only about seven minutes. Congratulations: now you know the magical elastic properties of "VR time."

But more important, now that you've gone into the headset—and through the frame that has always separated us from our stories—I'm hoping it's a little easier to imagine the feelings that VR and presence can elicit. Because as magical as those first few experiences can be, things are about to get a whole lot more impressive. Come on, let's go.

# 2

# ALONE ON A MOUNTAINTOP

## HOW "IN HERE" HELPS

## "OUT THERE"

THE SAN FRANCISCO BAY AREA is known far and wide for its natural beauty; foothills and mountains ring much of the bay, offering sweeping views to the people lucky enough to live on them (or intrepid enough to hike or cycle to their favorite vantage points). No matter how palatial your hilltop estate or indefatigable your legs, though, chances are you're not going to be where I am right now: sitting high in the Marin Headlands, a thousand feet above the bay, with San Francisco's glittering skyline in front of me and splendor all around me. The Golden Gate Bridge is off to the right; to the left, the Berkeley hills undulate. No one's allowed up here at night, so it's just me and the stars.

Despite the fall night, it's warm enough that I'm comfortable in just a light shirt and jeans; no need for the fleece jacket that's a year-round necessity throughout most of the area. Even the winds that buffet the area have died down. Stillness is all around. I breathe in deeply through my nose, imagining the air as clean white vapor, filling my diaphragm. I pause, then exhale slowly and deeply through my mouth—and watch, dumbfounded, as a stream of multicolored diamond shapes plumes into the night air. Those are my deadlines, I realize: the pitch meeting at work tomorrow that I don't think I'm ready for; my guilt at not calling my mom; that awkward interview from last week that I keep replaying in my head. The diamonds twinkle out like bonfire embers, and as my breathing continues, my anxiety begins to twinkle out too. Within minutes, I feel as serene as the night I'm sitting in.

"Okay," Chris Smith says from somewhere behind me. "Give us a sec, and let's go to the beach." I take the Oculus Rift off my head, and look around. In reality, I'm still high up—but I'm not alone in a mountaintop monastery. I'm in the "Consciousness Hacking mansion," a house in the hills of Silicon Valley. This is where Smith and his creative partner, Eric Levin, are living when I meet them; it's also where they work, coding away in a workroom strewn with computer cables and sensors. All around the Bay Area, in fact, people are trying to marry technology to New Age philosophy and finally crack that whole human-enlightenment thing. If they can pull this off, then VR might just prove to be the single most significant contributor to inner peace since Snoop Dogg's favorite plant.

## TOTAL COSMIC UNIVERSAL CONSCIOUSNESS—
## NO PSILOCYBIN REQUIRED

In person, Chris Smith and Eric Levin are more regular-looking versions of the *Flight of the Conchords* guys. Smith is shorter, with dark curly hair and a smattering of stubble. Levin is tall and lanky, bespectacled. And on paper, they don't seem like two people whose paths would naturally cross. Smith grew up on a farm in a small Methodist town in eastern Tennessee; Levin went to school in St. Louis and then spent time in Utah. Yet, each came to meditation and spirituality in ways that sound familiar to anyone who's whoa-manned their way through a late-night dorm room discussion.

In college, Smith was a Napster user—remember the 2000s, kids?—and through the music-sharing site he stumbled across some "binaural beats meditations," soundscapes that create the illusion of a beat by playing differing tones into each ear. (Proponents say that by coordinating the hemispheres of your brain, binaural beats can do everything from improve memory to lessen anxiety, but science has been less kind to such theories.) After that, a teacher's assistant got Smith into meditation, a practice that intensified when he moved to San Francisco a few years after college. Meanwhile, also in college, Levin had experienced a similar awakening, if a bit more dramatically: "The most powerful psilocybin experience I've ever had," he calls it, "just total cosmic universal consciousness."

He didn't know how to process the experience, so he started reading books about meditation. However, he was also prone to depression and anxiety, and two major episodes—once as an undergraduate while working at a summer camp, the other in medical school a few years later—made him realize that meditation was the only thing that kept him from, as he says now, "totally going off the deep end." But that depressive crisis in medical school did more than just make Levin a believer in meditation; it also changed his professional trajectory. He decided to use technology to help people transform themselves, so he dropped out of med school and got his master's degree in computer science in Utah.

Smith and Levin were in their late twenties when VR resurfaced, and both were immediately smitten. Smith left his job at a start-up and began building prototypes to explore how VR could enhance meditation (and vice versa). Levin had previously spent a summer at a Buddhist monastery in France working on meditation visualizations and sensed that VR could be key to unlocking that, but he wasn't quite ready to strike out on his own. When they met through a mutual friend, though, it wasn't long before Levin followed Smith's example, and the two joined forces. "I felt like I was getting a beam from the universe," Levin says. He gave notice at his job—he was working at a VR company in San Francisco—and he and Smith began experimenting.

## VR: PUTTING THE "IT" IN "MEDITATION"

Stress might not always be a killer, but it's definitely a nuisance. Whether it's the result of job pressure, money, health, relationships, or even media overload, more than three-quarters of all people regularly experience physical manifestations of stress, and just about half lose sleep over it. It's little wonder, then, that stress management has become part of a larger wellness movement in our society over the past few decades. After the popularity of meditation and yoga exploded in the United States during the 1960s and 1970s, meditation's younger sibling wasn't far behind. By the mid-'90s, the term "mindfulness" had fully emerged; by simply being aware of the present moment, evangelists argued, you could harness the benefits of more stringent meditation practice and become calmer, more focused, and more productive. By the 2010s, mindfulness and meditation had become a cottage industry—not just for self-help books, but for mainstream America. Oprah Winfrey extolled the virtues of the spiritual teacher and writer Eckhart Tolle and his emphasis on "awareness"; marketing-savvy scholars like Deepak Chopra ushered the so-called New Age into a new age, championing the mind–body connection in a lucrative franchise of books, retreats, and apps.

While Chopra and his rumored $80 million net worth might seem to have all the hallmarks of a nineteenth-century medicine man peddling elixirs, mindfulness is far from an empty promise. Various studies have found that meditation and mindfulness training can help not just with anxiety disorders, but substance abuse, depression, and pain. In 2011, a study conducted at a Harvard Medical School–affiliated hospital found that an eight-week mindfulness-based stress-reduction course resulted in "significantly greater gray matter concentration" in various brain regions that correlate with learning, memory, and awareness.

Not surprisingly, famously rationalist Silicon Valley got in on the act. A conference called Wisdom 2.0 has become an annual draw for start-up founders and tech executives to learn about "living mindfully in the digital age." Google began offering its employees meditation classes with names like Neural Self-Hacking. The house where Smith and Levin live in the hills overlooking Silicon Valley is owned by a roboticist named Mikey Siegel, who in 2013 founded the organization Consciousness Hacking. ("Hacking" is still a very sexy term in tech-minded circles and is often used to convey an unorthodox approach—even by corporate titans like Facebook, which named the road encircling its headquarters Hacker Way.) Consciousness Hacking was motivated, in part, by the belief that "modern technology, driven by science, has an incredible (and largely unrealized) potential to support psychological, emotional, and spiritual well-being."

That well-being is exactly what Smith and Levin are going for when I visit them. "VR is potentially going to become a direct interface to the subconscious," Smith says. "Traditional meditation practices use visualization, and you have to sit for years to get to the point where your visualization is powerful enough to do something with it." In VR, though, that visualization is immediate; Smith and Levin think the medium can jump-start the journey to a meditative state. So the pair isn't just hoping to build out the monastery experience they showed me. They want to create a suite of tools and then put their work in the hands of therapists, who in turn can help their patients—patients who may not have a VR setup at home, but who can get a few moments of virtual respite during their weekly or biweekly appointments.

The process may not even need a therapist. Imagine that your office included a meditation room, or even just a multipurpose quiet room you could sign up to use for ten or twenty minutes. There's a headset in there, some sensors that can read your heart rate and breathing, and a VR activity designed to get you into a flow state far faster and easier than just sitting down to meditate.

That, to me, sounds like heaven. I'm no stranger to meditation, but even in my most consistent phases I find that it can be a struggle; if the plan is to sit for twenty minutes, then invariably I spend the first half of that time, if not more, pushing thoughts out of my head. When I *do* manage to get to a point of mindful clarity, it's often right at the very end of my session. All runway, no takeoff. If VR can shorten that taxi time and get me to cruising altitude, though, then you'd better hope you're not standing between me and that headset. (I know, not very enlightened. Sorry.)

For Smith and Levin the endgame doesn't necessarily involve VR at all. It's developing mastery over the gearshift in your brain—however you achieve it. "We want the technology to be something you experience, then come out of it having tools you can use day-to-day without the technology," Smith says.

## MIND, BODY, EMBODIMENT

Carrie Heeter's office looks exactly like what I imagine her brain looks like. Tucked off the unfinished basement of her house in San Francisco, its walls are covered with bulletin boards and print-outs: nature scenes, notes to herself, prints of Kandinsky paintings. A *lot* of prints of Kandinsky paintings. It's busy, but it's also focused. Much like her life.

Heeter is a longtime professor in Michigan State's media and information department. Despite having lived in San Francisco for thirty years, she conducts her classes online, leading discussions among students around the world. She's also a game designer—not like *Halo* or *Call of Duty*, though. She calls her specialty "serious games," meaning play experiences that have a social or educational impact.

While most of her design experience is through traditional computers, Heeter has a long history with virtual reality. In the early 1990s, she and her colleagues created experiences that used a twist on the technology, something people called "second-person VR." Rather than wearing a headset, participants would wear 3-D glasses and stand in front of a blue curtain, where a special camera would film them; when they looked at a monitor, they could see a 3-D virtual version of themselves in a virtual undersea world, with octopi swimming up and taking hold of "their" arms. It wasn't VR as we think of it today, but seeing the virtual version of yourself being touched by a virtual creature created an astonishing sense of presence: 76 percent of people who underwent the experience felt a physical response to what they were seeing.

Two decades later, the rebirth of VR drew Heeter back in an unexpected way. Around the same time, she had begun practicing a yoga-meditation hybrid that brought her relief from the symptoms of multiple sclerosis. Rather than focus on a specific place or instruction—*imagine that you're in a meadow,* say, or *focus on your breathing*—the practice relied on Heeter finding her own preferred environment and then linking that imaginary place with physical relaxation exercises. As the body relaxes, so does the breathing, and the mind calms. Over time, the practice forges a link between the mind and the body. "To use software language," Heeter says, "it's 'object-oriented meditation.'"

The better she became at feeling her breathing and really knowing what relaxing *felt* like, the better those processes became linked with the scenario she was meditating about. She was effectively learning to think about feeling—which is something that our brains don't naturally do without a bit of training.

As humans, we're thinking all the time. Even when we're relaxing, our brains are as active as they are when we're solving a math problem. What changes isn't that we're thinking, it's the *type* of things we're thinking about, and the parts of the brain that activate when we're having those thoughts. When we're not focused on any sort of an external task that demands active engagement, a system in the brain known as the "default-mode network" kicks in. The default-mode network's chief purpose, many scientists believe, is to allow us to focus just enough on the outside world to notice whether anything unexpected happens. That might sound useless now, but for a hunter-gatherer on the savannah thousands of years ago, chances are it came in handy.

However, the default-mode network also kicks in when we get self-reflective—when we remember things that have happened to us, when we think about the future, when we consider moral dilemmas. And that, Heeter says, is where trouble starts. "The default network in the brain in today's society is the wandering mind," she says. "We're ruminating about the past, and we're worrying about the future. Or maybe even planning for the future; there's *some* productive thinking. But in general, a wandering mind is an unhappy mind. And that's where we spend all of our waking time: not being aware of everything that we're experiencing in the moment."

Heeter's own meditation had already led her to design apps and studies that investigated meditation's ability to calm that wandering mind. In one of those studies, she and her colleagues created a six-week program of calming meditation for health-care professionals working in hospice and palliative care, people who deal with what Heeter calls "enormously stressful situations at work." In just six weeks, they found improvements in burnout and compassion fatigue—both common stress disorders found among caregivers.

When VR came back around, she realized that she had an opportunity to bring those benefits to even more people. "Lots of people aren't willing to sit for ten minutes and close their eyes," Heeter says. "But even looking out the window or looking at pictures of nature has anti-stress benefits." Maybe virtual reality could bring more people into meditating, she reasoned; even with their eyes open, they'd get some benefit. Her meditation teacher, a man named Marcel Allbritton, was also a therapist, and the two of them began wrestling with what VR meditation might look like—how it could help strengthen people's ability to feel their bodies, recognize their emotions, and turn off their wandering minds.

The answer, it turns out, is sitting on her computer. When I put on the headset she hands me, I find myself on a virtual rendition of Costa del Sol. Unlike the real Spanish tourism destination, this beach I have to myself. It's a stunner, too: palm trees dot the sand, and rock formations rise out of the water. I can hear waves and birds, and soon they're joined by a man's voice in my headphones. He's telling me to take in the scene, to look at and listen to everything around me. After giving me a chance to do that, he asks me to connect with my body. "Feel your hands," he says, "what you're sitting on." I wouldn't have noticed it without the prompt, but I realize that my elbows are on the chair's armrests, and one of my feet is tucked awkwardly underneath me. I adjust my body so that I'm sitting evenly with my hands in my lap.

Immediately, I feel my body relax through a tension I hadn't even registered—and when that tension dissolves, a tiny epiphany takes its place. If you're a yoga aficionado, or you start each day with a meditation, then you're likely used to taking stock of your body and relaxing. Not me. I've had terrible posture and flexibility my whole life, but I'm not aware of it; I don't notice how awkwardly I'm sitting as much as I later feel the effects of it. A single voice prompt, though, helped me to notice it, and then to correct it in a way that calmed my body and made it easier for me to calm my mind.

The key, Heeter says, is something called interoception. It's a term that's gained ground in psychology circles in recent years and basically means awareness of bodily sensations—like my noticing the fact that I was sitting awkwardly or that keeping my elbows on the chair's armrests was making my shoulders hunch slightly. Not surprisingly, mindfulness meditation seems to heighten interoception. And that's exactly how Heeter and Allbritton structured the meditation I'm doing on Costa del Sol. First, I connect with the environment; then with my body; then I combine the two. The combination of VR and interoception leads to what she describes as "embodied presence": not only do you feel like you're in a VR environment, but because you've consciously worked to integrate your bodily sensations *into* VR, it's a fuller, more vivid version of presence.

And that embodied presence, it turns out, is a two-way street. The weekend after my visit with Heeter, my wife and I go hiking in the wooded hills above Oakland, near where we live. It's an activity we picked up when we were living in New York City, and it continues to be one of the best ways I've found to clear my head and feel grounded. This particular trail is high on a ridgetop, with stunning views of San Francisco; on a balmy January afternoon it's hard to imagine anything more peaceful. I stop to look around. To my left, there's a grassy hillock; to my right a stand of birch trees extends down the hill. Around me, leaves rustle in the breeze. The sun is well on its way toward the horizon, but I turn my face upward to take in the blue. Wispy cirrus clouds sweep up from the east. It's perfect. Yet, I keep scanning, seeking … *something*. What that something is, I don't know—until all of a sudden I do. I'm looking, I realize, for pixels.

## BABY GOT BIOFEEDBACK

Like most people who are trying to figure out how to make VR a vessel of self-discovery, Josh Farkas practices what he programs. When he left his graphic design job a decade ago to start his own company, he was plagued by doubts, and "after failing dozens of times" to deal with the stress, he discovered meditation and mindfulness. "It was this pivot point in my life," he says, "of being able to get things under control."

When he tried to get other people excited about meditation, though, Farkas didn't have much luck. *Well, that's not for me,* they'd say, or they'd feel weird about the suggestion. "What's a bummer about meditation generally is that the people who can do it the best are the ones who need it the least," he says. "If you can go to a special place within yourself, you're already doing pretty good—it's the people who aren't able to do that who need it."

So when VR came along, Farkas, now thirty-four, realized that it might be the perfect way to get people to see the light. The result is Guided Meditation VR (GMVR), a . . . well, a guided meditation app. When you begin, you choose one of a dozen environments that range from a temple in the Hana Valley to desert canyons to outer space to Costa del Sol. (Farkas is a fan of Carrie Heeter's work, and gave her a customized version of the environment to develop her own meditation.) You can opt for a stationary experience or for one where you feel like you're flying slowly through the landscape. Next, you choose what *kind* of meditation you want: Do you want to find relaxation? Compassion? Mindfulness? Then you pick music, and how long you want the meditation to be, then your current mood. There are thousands of permutations; as the menus keep branching, your experience becomes more and more personalized.

Then comes the interesting part. If you're using the smartphone version of GMVR, it uses the phone's front-facing camera to read the heart rate in your fingertip before and after your meditation; if you're using the high-end PC version, it listens to your breathing through the built-in microphone. The same thing happens when your meditation finishes. (The camera and microphone aren't as accurate as a dedicated heart-rate monitor or breath sensor, but they're enough to give you a general sense of your stress level.) In other words, you don't just feel the immediate effect that meditating has on your mood. You *see* it.

Now, when Farkas goes to conferences to speak about the potential for VR, he says, people sometimes come up and hug him. He's heard from a man who credited his app for helping him get his first night's sleep since his child died. He's heard from an elementary school student in Australia who used it to feel less anxious about an upcoming surgery. *Every time I get nervous now,* the kid wrote, *I think about Japan, and I think about sitting there and looking at the river.* "That gives me chills," Farkas says. It's bigger than just people at home, though. Farkas sees a world where VR's stress-reduction properties make VR a staple in dentists' offices and medical settings.

In fact, apps like GMVR are the very first step toward using VR to help you become aware of your own physiological processes— and then learn to adjust them. "Biofeedback," as it's known, is like a training regimen for your brain-body connection. Studies have found biofeedback to be effective for physical ailments as pedestrian as motion sickness and urinary incontinence, but it's also been the focus of plentiful research as a tool to help treat anxiety. Heart rate, as you can probably guess, is a huge part of that. If you've tried a heart-rate app on your phone, then you're likely familiar with the weird, almost therapeutic effect it can have: if you watch the readout while it's measuring your heart rate, you may be able to lower it by willing yourself to breathe more deeply. (Granted, on the flip side, you also might get anxious about your heart rate being high, thus forcing it higher.) The more you do it, the more familiar you become with the sensations and cues connected with lowering your heart rate, and the easier it becomes to do—just like meditation.

It's not hard to see where this is going. Simple heart rate is just the beginning. As sensors get smaller and more powerful, headsets will be able to monitor not just breath, but heart rate variability— the ever-shifting periods between each heartbeat. (The higher your heart-rate variability, generally, the better off you are.) And that data can be turned into visual cues—cues that can help even beginners see some palpable benefits. One example: instead of just imagining your breath changing color as you mindfully try to lower your heart rate through breathing, you could actually see it happen. Something that can otherwise take years of practice is now, conceivably, minutes away. And when, as in Chris Smith and Eric Levin's prototype, it's already possible to sit on a mountaintop and breathe a plume of glittery particles into the night air, that kind of breakthrough feels like it's just around the corner. One headset appeared at an electronics trade show in January that had EEG sensors already built in.

Just because things are moving fast, though, doesn't mean that everything sticks. A few months after my visit to the Consciousness Hacking mansion, I email Smith to see whether I can come back and see what kind of progress they're making. "Lots of new-

54

ness on our end," he writes back excitedly. It turns out that Levin has headed to South America to "pursue the spiritual path more deeply," and Smith is working on another project. This isn't a total surprise; for all the frenzied development happening in VR, small start-ups are still prone to flux. It's the first of many strings that will resist getting neatly tied up during my travels through VR. My path to virtual enlightenment will have to continue without their help.

## "IN HERE," "OUT THERE," AND THE TRICK TO COEXISTENCE

Whether in a mansion, a cluttered office, or my own house, all of this VR has had one thing in common: it's me, by myself. That's to be expected; meditation tends to be an inwardly focused practice. But virtual reality doesn't just promise to revolutionize that self-connection through monastic seclusion—or so I realize when I look at the clock in Ray McClure's workshop. He and I have been . . . well, there's really no easy way to say this: we've been grunting and cooing at each other for nearly a half hour.

McClure is an artist-in-residence at the Gray Area Foundation, a digital incubator program housed inside a historic movie theater in San Francisco's Mission District. (And yes, that is just about the most Bay Area sentence you could possibly write.) He's forty, with a thin mustache and a conversational pattern that's as energetic as it is syncopated—like a pull-string motor that refuses to catch. More than a decade ago, he became one of Twitter's first employees, with stock options that enabled him to buy a house and open an art gallery in the city. (Okay, maybe *that's* the most Bay Area sentence you could possibly write.) Now, he and his creative partner have created a VR experience that allows you to make spontaneous, unconventional, sometimes uncomfortable art with other people.

VVVR, short for "visual voice virtual reality," is "a new social system that blends fantasy and reality to foster ego-free communication with real people, in a space beyond words." It's not meditation, though the two share some obvious similarities—a fact that becomes apparent when Ray and I sit down on cushions facing each other and put on our headsets.

The first thing I notice is that we're both there, sitting cross-legged, about six feet apart, just as in real life, but we're surrounded by nothing but white. The second thing is that we're both bald, blue, and wearing nothing but flowing robes. The third is that when he starts talking to me to explain what to do, a stream of cubes and spheres starts flying out of his mouth. In VVVR, your voice is the controller, and all you need to do is use it to create. The type of sound, as well as its pitch—whether a sonorous *ummmmmm,* a staccato *kuh!,* or a high-pitched *eeeeee*—determines the shape, and color, of the forms that sail out of my mouth. Some are orange and spiky, others are rounded and green; it's only through experimenting that I can start to figure out how to make, say, a pyramid. But in order to do that, I have to do something important: get over myself.

Even now, two or three years after virtual reality became easily accessible, self-consciousness might be the greatest threat facing the technology. Not ergonomics, not simulator sickness, not price. Self-consciousness isn't even a problem a company can solve—it's completely dependent on the person wearing the headset. Nor is it an obvious sort of problem; self-consciousness doesn't stop people from playing a VR game in their own homes, or watching a movie in VR "with" someone else. It's really only a factor when using VR around other people.

Some of that awkwardness makes sense, depending on your surroundings. You wouldn't put a blindfold on while riding the subway, right? Why put yourself in a similarly compromised situation for VR? (Air travel is the big exception here; just as we wear sleeping masks on planes, VR headsets are becoming more and more commonplace for people who want to spend their flight watching movies or browsing the web without reminders that they're sitting in a tube full of strangers seven miles above the ground.) Some of it is also a function of what headsets look like in these early days. Just as the ill-fated Google Glass immediately stigmatized all its wearers as "glassholes"—aka "techier-than-thou douchebags who dropped fifteen hundred dollars to see an email notification appear in front of their face"—so too do some VR headsets still look like face TVs for nerds. Why else do you think you see people wearing them in commercials but not in coffee shops?

Safety and aesthetics aside, though, a more insidious force is at play, which sociologically speaking is probably something about "adherence to social norms" but is really the fear of doing something weird. Think about it: VR is all about you experiencing a different reality than everyone else around you. That means that however you're behaving because of VR—gasping in surprise as a *T. rex* appears at the end of a hallway, say, or crouching down in order to examine something on the ground—is going to be out of step with what other people are doing. Think you're brimming with self-confidence? Proud that you don't care what other people think of you? You haven't been asked to sit in a public place with a VR headset on and create sustained nonsense.

Even sitting with McClure in his private office, it's difficult to do. "AhhhhhhhhwwwwwwhaaaaAAAAAHHHHHHHHH," I start, opening my mouth wider and wider as if for some invisible but demanding dentist. I can hear my voice in my headphones; while it's pitch-shifted to sound a little spacier than usual, it's still recognizably mine. To be candid about it, I feel like a complete ass. But making stream-of-consciousness noises is much easier when there's a literal stream of consciousness issuing out of my mouth in VR. I turn my head from left to right, watching the shapes and colors fan out as I vocalize. "Raahhhhhhhhh. Nnnngggggaaaah. Huhhhhhhhh. Gah gah gah gah NYAAAHHHHHH"—I'm in falsetto now—"AhhhHHHHhhhh. Yih. Yih. Howwwwshhhhhh. Mamamamamamoh."

Honestly, the only way I can describe these sounds is by later listening to a recording I made at the time. In the moment, though, I'm not thinking about the sounds I'm making; I'm thinking about the colors and shapes I'm making—and the ones that bald blue McClure is making, sitting across from me. Larger, diffuse clouds of color begin to drift by overhead, lending a touch of ambiance to our otherwise minimal surroundings.

Slowly, I realize, my focus has shifted. When I first put the headset on, I was keenly aware of being in two places at once: in here and *out there*. I was experiencing presence in VR, certainly, but it was tempered by wondering what I looked like to someone out there. (And, let's face it, what I was doing was pretty out-there in more ways than one.) But as McClure and I created a visual language together, that self-consciousness ebbed, and so did my feeling of straddling a "here" and a "there." Now, I'm just here. And just as important—and believe me, I realize how this sounds—here, I'm just now.

I'm not the only one engaged by the experience. McClure and his partner have taken VVVR to a few gatherings, among them a secret invite-only event hosted by movie director David Lynch called the Festival of Disruption. Over the course of a weekend, they put hundreds of people inside VVVR, two people at a time. Actor Owen Wilson loved it; so did Robert Plant's band. McClure plays me a video montage they made of the event: it's person after person after person losing themselves in the sounds they and their counterpart are making. There's no trace of awkwardness, despite the fact that a roomful of people is in eye- and earshot. To paraphrase Funkadelic, it's a clear case of freeing your mind so your voice will follow.

Eventually, I want to ask McClure a question, but it feels wrong to pollute our sound world with actual words. Besides, it feels like it's been only five minutes in there, maybe ten at the most. I take off the headset and look up at the clock. Twenty-three minutes. That's VR time for you—I'm late for a meeting back at work.

When I leave the Gray Area theater and hail a Lyft ride back to the office, I notice that the driver has stuffed the pockets behind the front seats with laminated photos. They're all selfies that she's taken with celebrities who have gotten in her car. There's an R&B singer, a football player, an actor. And my first reaction, I have to admit, isn't great. I start to inwardly scoff. *Why would you do that?* I think. *Why are you trying to impress people with something that happened by random chance?* But then, almost immediately, I become aware of what I'm actually doing: judging her for no reason at all, for doing something that makes her happy.

I think back to moments before, when I was sitting with Ray McClure, droning nonsense. In that space, I was thinking about nothing but the thing I was doing. I didn't feel judgment, or expectations, or worry about how I might look or sound. Which is how everyone should be able to live their lives. So I lean forward between the seats of the car. "Tell me about the nicest celebrity you ever picked up," I say to the driver.

# 3

# HEDGEHOG LOVE

## ENGINEERING FEELINGS
## WITH SOCIAL PRESENCE

I N JANUARY OF 2015, consumer headsets were more than a year away—yet, the Sundance Film Festival was experiencing the onset of VR fever. The festival's New Frontier program, which celebrates "the convergence of film, art, media, live performance, music and technology," featured thirteen installations, eight of which were virtual reality. They ranged from a short film about refugee children (*Project Syria*) to a short campy ode to Japanese monster movies (*Kaiju Fury!*) to a short thought-provoking piece about date rape (*Perspective, Chapter 1*) to a short documentary about VR itself (*Zero Point*). All were, yes, short. But more important, each was different from all the rest and thus took small steps in its own direction toward figuring out the new rules of whatever VR filmmaking would look like.

That year would prove to be an inflection point; the festival would soon establish a stand-alone VR program to allow for the crush of submissions it was getting. Looking back with the benefit of hindsight, though, the most important piece of VR at Sundance wasn't a festival selection. It wasn't even a completed work that you could actually *watch*. It was a title and premise only, mentioned in passing as part of an announcement that Oculus would be producing animated shorts. *Henry*, the company said, would be a comedy about a hedgehog who loves balloons.

It was that, but despite being a mere twelve minutes long, it was also much, much more.

## THE HEDGEHOG
## WHO COULDN'T HUG

When *Henry* begins, you're sitting in an apartment that's basically the classic six of cartoon-animal real estate: part Ewok village, part artist's cottage. (Really, what's an Ewok but a hedgehog with a spear?) Picture frames line the walls, and a slice of a stout tree branch in the middle of the floor serves as a coffee table. Behind you, an easy chair sits next to a wood-burning stove. There's a teakettle on the stove, and a stack of newspapers next to the chair.

It's no ordinary day in Henry's home, though. Above you, leaves strung together next to a cluster of floating balloon animals have HAPPY BIRTHDAY written on them. Noises off to your left alert you that someone is puttering around in the kitchen. If you crane forward to get a better view, you see Henry, absentmindedly talking to himself in adorably high-pitched hedgehog-ese. Finally, Henry walks out, holding a feast: a single strawberry on a tray, capped with a dollop of whipped cream. It's all unbearably endearing, like you walked through a movie screen and woke up in a Pixar film.

If this is a Pixar movie, though, it's *WALL-E*. Because from the moment Henry puts a single candle in the strawberry, you're not entertained as much as you're devastated. It's *his* birthday, you realize, and he has no one to celebrate with. He puts on a brave face, tossing some confetti in the air and blowing a tiny noisemaker, but not even the ladybug crawling across the table wants to stick around. And when his eyes slide in your direction, his giggles giving way to a sad sigh, all you want to do is to take him home with you.

But why? Some of it is because of smart character design, clearly informed by decades of Disney movies: his eyes are huge and his birthday wish universally heartbreaking ("I want a fwwiiieeeeend!" he whimper-whispers fervently). Some of it is

When Henry's eyes meet yours, you're no longer an audience member—you're there with him.

irony-rich writing: when his wish comes true, and the balloon animals come to life, they're terrified of his spines and flee from the hugs he so desperately wants to give them. But above all, it's the fact that despite your invisibility, you're there in a very important way. You're not just a witness. You're an attendant.

If that doesn't quite make sense, think about what *Henry* would look like if it were a regular, screen-contained, see-it-in-the-theater movie. (Also, don't get too sad worrying about Henry. No spoilers, but everything works out in a properly heartwarming way.) You'd see his apartment, but you'd see only the things that a screenwriter and director had deemed necessary to the story, and only in the order that an editor had decided made the most sense. You'd watch Henry put the candle on his strawberry and blow his tiny noisemaker, but when he felt a pang of loneliness, he wouldn't look at the camera—he'd just stare wistfully out the window. And when the balloon animals came to life, instead of watching Henry's reaction devolve from wonder to excitement to despondence, you'd instead get a slapstick barrage of quick cuts as he chased them around his apartment. Every emotional beat, in other words, would be choreographed by someone else, and presented to you.

But in VR, you're there. You're there the whole time, in sequence, without interruption. You see everything Henry sees and feel everything he feels. And, perhaps most important, he sees you. His eyes lock on yours; he acknowledges your existence. He'd be breaking the fourth wall—if you hadn't already clambered over it to get into his apartment with him.

Remember presence? This is the beginning of *social* presence. Mindfulness is cool, but making eye contact with Henry is the first step into the future.

## SOCIAL PRESENCE:
## THE SEED OF SHARED EXPERIENCE

Back in 1992, our friend Carrie Heeter posited that presence—the sensation that you're really *there* in VR—had three dimensions. There was personal presence, environmental presence, and social presence, which she basically defined as being around other people who register your existence:

> [I]f other people are in the virtual world, that is more evidence that the world exists. If they ignore you, you begin to question your own existence. The Hollywood fantasy theme of a human who becomes invisible to the rest of the world, and is able to move freely around (and through) people, exemplifies the experience of reduced presence in that kind of hypothetical virtual world. However, if the others recognize you as being in the virtual world with them and interact with you, that offers further evidence that you exist.

Just as being able to move in a virtual world makes it more real, so does being acknowledged in a virtual world. And when Henry looks at you, that's exactly what happens. Of course, when people were thinking big thoughts about virtual reality twenty-five years ago, they weren't necessarily considering the possibility of a hyper-cute hedgehog, but Heeter left room for presence involving nonhumans: "Social presence can also be created through computer generated beings."

The fact that this happens during *Henry* is all the more amazing because you're not exactly there. I don't mean you're in VR and so there *is* no "there"; I mean that if you look down at yourself, you don't see a body—no hands, no legs, no evidence that you exist in the world of the story. Nor do you have a way to interact with Henry; even if you talk, he won't hear you. You're basically the shyest party guest of all time. (But possibly also the best party guest of all time.) There's nothing social about Henry, other than the fact that his eyes meet yours.

And even that isn't unique. The entire concept of breaking the fourth wall hinges on a fictional character crossing the divide between fiction and reality and addressing the viewer directly. Since Oliver Hardy first shot an exasperated look at the camera in the 1920s, characters in movies and TV have cribbed the move. Think back to Ferris Bueller in *Ferris Bueller's Day Off*. Amélie in *Amélie*. Deadpool in *Deadpool*. Notice a trend? But the phenomenon isn't restricted to titular characters. In *Annie Hall*, for instance, Woody Allen treated the fourth wall like the mound of cocaine his character sneezed into oblivion.

Again, though, consider how these acts play in a conventional 2-D movie. Sometimes they're surprising; often, they're funny. Seldom do they provoke a serious emotion or create a connection between the character and the viewer. In *Henry*, however, those moments of eye contact have an impact that far outpaces any vaudevillian mug or superhero metawisecrack.

So what exactly happens in those moments? It depends. Shared eye contact, or what social scientists call "mutual gaze," can have some pretty stark effects. In 1989, three psychologists decided to test a premise that had been around since Charles Darwin: the idea that emotion can be not a cause of behavior, as we so often assume, but a *result* of it.

To do this, they rounded up almost one hundred college undergraduates and then randomly matched forty-eight pairs of male and female subjects—first establishing that they didn't know each other—and put them in a test room. They gave each person one of three instructions to follow for two minutes:

1. Look at the other person's hands.

2. Count the other person's eye blinks.

3. Look into the other person's eyes.

After the two minutes were up, the man and woman were taken to different rooms, and each filled out a questionnaire based on something called the Rubin Love Scale (no relation, unfortunately). Given the instructions, there were five ways things could have played out, only one of which resulted in mutual gaze. Just as the psychologists hoped, though, the pairs who had gazed into each other's eyes expressed significantly more affection and respect for each other than any of the other volunteers did for their own counterparts. All well and good—but the experimenters thought that Rubin's Love Scale was a little limited (hey!), so they ran a second study.

The Rubin scale, they wrote, measured "dispositional love"—whether someone was likely to forgive their partner, for example. It measured the probability of certain behavior rather than emotion itself. The researchers wanted to see whether mutual gaze between strangers could give rise to "passionate love," one based more on signs of physiological arousal. So this time, they put together thirty-six pairs of male and female strangers and told them they were taking part in an experiment about extrasensory perception. Before it began, each volunteer filled out a questionnaire that contained not just questions based on Rubin's scale, but also some drawn from clinical discussions of passionate love, as well as some based on in-depth interviews with actual couples. For example, they had to rate how strongly they agreed or disagreed with the statement, "When I see _____, I feel excited."

For the "ESP test"—this was 1989, *Ghostbusters II* was in theaters—the volunteers were again brought into a room and asked to look either at the other person's hands or into their eyes. This time, though, the room could be either lit normally or with the lights turned low and jazz piano playing. After two minutes of that, the volunteers were asked to perform a second ESP test, one in which they performed exaggerated smiles and frowns while their counterpart tried to describe the "symbol" the person was sending. Then they filled out new questionnaires.

As expected, mutual gaze resulted in increased passionate love, as did jazz piano and dim lighting. However, there was a bit of a surprise as well: the "romantic setting" had a significant effect only on people who had proved to be emotionally susceptible to their own facial expressions. In other words, the people who felt happier after smiling or angrier after frowning were more affected by the lighting and music. For them, a number of different behaviors could give rise to emotion. But even with the others, prolonged eye contact stirred the pot of emotional connection.

So what does all this mean? Well, it means that not everyone is going to get turned on by a smooth jazz album, like they did in the study. It means that people have different buttons; some are simply more suggestible to outside cues. But it also means that the eyes aren't just the windows to the soul, but to the heart as well. And even to the libido.

Before you start composing that angry email: I'm not suggesting that you should be cultivating an amorous attachment to a hedgehog. I'm just saying that something as incidental as eye contact can have a real effect. When presence is a factor, like in an immersive VR environment, that effect can go from real to profound. Feelings, it seems, can be engineered—even for a CGI animal. And if VR can forge an emotional connection between you and *Erinaceus concolor,* imagine what it can do with a CGI person. Or even a *real* person. That whole windows-to-the-soul thing takes on even more weight in VR, where you're already primed to connect.

## FROM PASSIVE TO MASSIVE(LY IMPORTANT)

*Henry*'s impact wasn't exclusively inside the headset, either: in 2016 the short became the first piece of original VR content to win an Emmy. But while its statuette says "Outstanding Creative Achievement in Interactive Media—Original Interactive Program," its interactivity is somewhat limited. When you visit Henry's apartment, you're there to watch, not to participate.

Regardless, VR had won a television award for something that was part cartoon, part video game, and completely unprecedented. The industry was growing fast, with old Hollywood studios and VR-first creative companies converging on this undefined terrain, and the need to agree on some common terminology became a concern. Simply "VR," after all, was too fuzzy; it could mean anything happening in a headset, from a video game to a meditation environment. "Filmmaking," for its part, is a term born of a now-archaic process; what do you call it when there's not only no film, but no frame? In the search for a just-right term, creators defaulted to the most general term possible that could still mean something—"storytelling." And shorts like *Henry* came to be known as "experiences." It made sense: you kinda watched them, and you kinda played them, but you definitely experienced them.

As VR storytelling grew and VR experiences proliferated, *Henry* was joined by other CGI shorts. *The Rose and I,* a *Little Prince*–inspired experience in which a young boy finds an unexpected friend, came out of Penrose Studios, a storytelling company founded by an Oculus alum named Eugene Chung. That same year, Baobab Studios—a VR start-up cofounded by the director/screenwriter of the Madagascar animated films—released *Invasion!,* a charming comedy short about a huge-eyed bunny defending Earth against alien interlopers.

Both played at film festivals, both received critical acclaim, and both were just the beginning. By now Oculus Story Studio, Penrose, and Baobab have all released multiple experiences, some of which have taken steps in entirely new directions. (However, Oculus Story Studio is no more, having ceased operations in 2017. "Now that a large community of filmmakers and developers are committed to the narrative VR art form, we're going to focus on funding and supporting their content," wrote Jason Rubin, an Oculus executive.) VR experiences are no longer Pixar clones eliciting "awwwww" reactions. Some are meditations on grief and love: in Oculus Story Studio's breathtaking *Dear Angelica,* a young girl reads letters from her late mother, an actor, while the woman's movie performances thunder through the virtual space. Others, like Penrose's *Allumette,* present you with a floating city, one you can peek into like a VR diorama to follow the melancholy story playing out. (Just as *The Rose and I* has a *Little Prince* feel, "The Little Match Girl" is *Allumette*'s spiritual ancestor.)

We've seen this experimentation before—when film first emerged more than a century ago. Just as the Lumière brothers created the illusion of a train rushing at the screen in 1895's *Arrival of a Train at La Ciotat*, or 1902's *A Trip to the Moon* explored how to tell a story through multiple scenes, so too does each early VR experience pioneer new storytelling techniques. It took decades for film to create a visual grammar: cuts, reverse angles, and montages may all be familiar to moviegoers now, but once upon a time each was just a director's wild stab at conveying information inside the constraints of a new medium.

Now that VR brings you inside the frame, though, those constraints are gone—and it's time for a new generation of storytellers to try a new generation of narrative techniques. Many of those techniques are going to grapple with directing your attention; VR experiences don't have the benefit of a rectangle to bound your focus, so creators need to find ways to make you notice the things that matter in the 360-degree sphere that is the VR "screen."

But the most interesting new techniques will continue *Henry*'s tradition of forging a connection between you and the characters. In *Invasion!*, for example, if you look down at your own body, you realize that you're not just watching a bunny—you're a bunny too. That changes the dynamic between you and the character. At one point in the experience, the bunny does a tiny dance; when *Invasion!* was first screened at film festivals, people wearing headsets tried to dance with the bunny, caught in a rare moment of species kinship.

However, one common thread persists through the early VR experiences: you exist within the frame, and you might even be acknowledged by characters, but you have no agency. The story can interact with you—not the other way around. So the question becomes: Is there a way to bring you into a virtual experience to increase that social presence, so that you feel even more a part of the fictional world?

There might be—and it's already in our hands. But it's going to take a bit of explanation first.

## MANUAL OVERRIDE: THE ROLE OF HAND PRESENCE

In Chapter 1, we explained the difference between mobile VR and PC-driven VR. The former is cheaper and easier; all you do is drop your smartphone into a headset, and it provides just about everything you need. Dedicated VR headsets rely on the stronger processors of desktop PCs and game consoles, so they can provide a more robust sense of presence—usually at the cost of being tethered to your computer with cables. (And also at the cost of actual money: dedicated headset systems run hundreds of dollars, while mobile headsets like Samsung's Gear VR or Google's Daydream View can be had for mere tens of dollars.)

There's one other fundamental distinction between mobile VR and high-end VR, though, and that's what you do with your hands—how you input your desires. In the world of video games, input happens via controllers, which can range from simple joysticks to Xbox gamepads to ultra-complicated keyboards. Even plastic steering wheels and guitars can be input devices, if you're playing a driving game or *Rock Band*. When VR reemerged in the early 2010s, however, the question of input was open to debate. Actually, more than one debate.

Part of the problem was practical. If you had a headset on, you wouldn't be able to see what you were using with your hands, so you needed a device that was intuitive. In other words, a keyboard and mouse wouldn't work. Even a conventional game controller might be too complicated. But the more interesting part of the input conversation revolved around extending the sense of presence. When you put on a VR headset, your head effectively becomes the center of a virtual space—but was there a way to create *hand* presence as well? In other words, could you bring your hands into VR?

Video game controllers are basically metaphors. Some, like steering wheels or pilot flight sticks, might look like the thing they're supposed to be, but at their essence they're all just collections of buttons. When you press those buttons, that input is translated into some sort of in-game action, whether it's "grab" or "honk" or "open." But if you could somehow do away with that metaphorical layer, or at least create a new one that felt more natural, you could make someone feel as though the hands they see in VR are their own.

The building blocks of hand presence already existed in various forms. The Nintendo Wii video game console had become a worldwide phenomenon in part because it used not a complicated gamepad, but handheld wands that were tracked in space: the console knew how you were holding the "Wiimote" and what you were doing with it, which allowed you to use it as a sword or a bowling ball or anything else that a game developer could imagine. If you weren't careful, you could even use it to destroy your flatscreen TV in real life, as plenty of people found out.

An ocean away from Japan, Microsoft had created an infrared camera called the Kinect that worked with Xboxes. A Kinect could scan the room you were standing in, identify you as a person, and track your body and hands so that they could work as controllers, or even be rendered in a game. (It didn't always work perfectly, but that didn't stop companies from making lackluster games for it.) And third, a company called Leap Motion had created a sensor that was essentially a more focused version of the Kinect. It didn't try to track your entire skeleton, but by just concentrating on your hands, it was able to track them with remarkable precision, down to the smallest finger waggle. The dream use for Leap Motion was being able to use your hands, instead of a mouse, to control your computer; a few computer companies licensed Leap Motion's technology, but no one was re-creating any scenes from *Minority Report.*

Thanks to those kinds of conceptual predecessors, creating a very basic motion-trackable controller was actually easy. Now, both of the major mobile VR headsets come with a small remote-control device that's essentially a tiny version of a Nintendo Wii controller. When you're in VR, you can use it as a laser pointer to select and navigate experiences; as a game controller, it can become a fishing rod or flashlight.

But although those remote-style controllers let you use your hands, they don't let you *have* your hands. For that, you need to use controllers that work with more powerful desktop VR systems. These do a few very important things:

1. They render a simulation of your hand in VR that is based on the controllers' position and orientation. It's not done visually, so your virtual hands won't be wearing your rings or even have the exact shape or skin tone as your own hands—but your brain is pretty easily fooled in this respect. Numerous studies have shown that your sense of "body ownership" can easily transfer to a virtual version of a hand, even if that hand looks markedly different

from yours. In fact, your perception of your own virtual body can affect your real-world behavior: in one study, volunteers who were given an overweight virtual body moved their head more slowly than those given an underweight virtual body. (We'll discuss these phenomena later in the book.)

2. These controllers are completely intuitive—you can give them to someone who has never played video games, and after a brief tutorial that person will know how to handle them. The foundation of this simplicity is having buttons that are placed on the controller to match up with how you might use your hand in real life. For example, in order to pick up an object in VR using the HTC Vive's wand-shaped controller, you use a trigger-like button on the rear of the controller; the very action of pulling the trigger closes your hand, so the real-life motion matches what your hand is doing in VR.

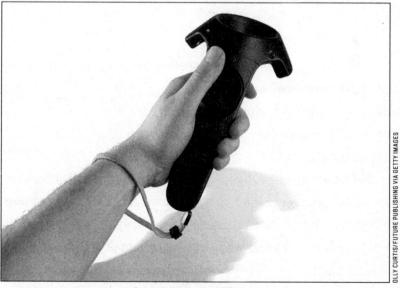

OLLY CURTIS/FUTURE PUBLISHING VIA GETTY IMAGES

A controller for the HTC Vive.

3. They bring your real-life hand movements into VR. The buttons on Oculus's Touch controller, for example, are capacitive: like a touch screen, they know when your fingers are contacting them. If you execute simple motions—waving, pointing at something, or giving a thumbs-up—the controller translates those into a similar gesture for your virtual hands.

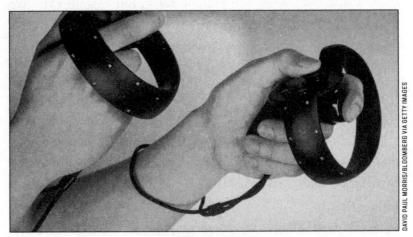

Oculus Touch controllers.

When all these things come together—flawless tracking, intuitive "user interface," an ergonomic design that allows the hand to assume a natural position, and gesture translation—hand presence becomes possible. Now it's not just your head that's in VR, but your hands as well.

Before we move on, let me point one thing out: this is just the so-called first generation of VR controllers. The next iteration of the HTC Vive's controller won't be actively held at all but will strap around your hand, allowing you to hold and drop objects by closing and opening your hands. Last year, Leap Motion shrank its finger-tracking sensor down to something so small it could be embedded in any smartphone or VR headset. Trying it for the first time, I couldn't stop moving my fingers around pinching small virtual boxes to stack them in a tower. Playing with blocks has never been so riveting.

In other words, any VR system could conceivably do away with controllers completely for certain tasks, finally delivering on that *Minority Report* promise. Hand presence isn't even necessarily limited to hands: HTC sells small wearable trackers that you can affix to any object, or any body part, to bring it into the Vive's VR. People have attached them to Nerf guns, tennis rackets, even to toy baseball bats to take virtual batting practice. They've attached trackers to their cat so that they don't accidentally step on poor Sir Flufferson during a spirited VR session. And by attaching enough to their hands, feet, and joints, people have had surprisingly lifelike dance parties in VR. None of this is easy—they're all just experiments by developers—but it gives you a glimpse of how you can "virtualize" just about any real-world object. (Later in the book, we'll visit a company that's using a similar idea to create a next-level version of laser tag.)

In 2017, Mark Zuckerberg posted to Facebook a photo album of his visit to Oculus's super-secret research lab outside Seattle. This was a tiny bit upsetting, if only because I've been trying to get in there for years, begging their gatekeepers to no avail—but fine, I get it, he's the CEO. One of the pictures showed him wearing a headset and a huge grin on his face, and some thin white gloves on his hands, one of which was making a gesture immediately familiar to any comic-book fan. "We're working on new ways to bring your hands in virtual and augmented reality," he wrote in the caption. "Wearing these gloves, you can draw, type on a virtual keyboard, and even shoot webs like Spider Man."

First off, Mark, it's *Spider-Man*. Nerd demerits for you. Second, being a webslinger isn't on everyone's wish list. But being able to use your hands freely in VR opens up a huge new set of possibilities—possibilities that will turn VR from something you use for entertainment to something you use for just about any communication you can imagine. It sweeps away more than just the clumsy metaphors of game controllers; it removes every obstacle between your body and the virtual world. And when your body is there, it's that much easier for your emotions to be there as well.

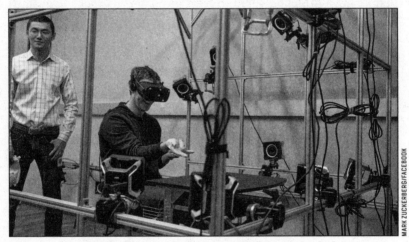

Mark Zuckerberg taking hand presence to the next level.

MARK ZUCKERBERG/FACEBOOK

## WAIT A SECOND—
## YOU WERE TALKING ABOUT STORYTELLING

Great point! That's exactly what I was doing. But now you understand what hands can do in VR, and what they might soon be able to do in VR. That second part matters because *Henry* and all those other VR experiences were created to be consumed and navigated with no controller at all. Any social presence that you experience while watching them comes in small doses. That's changing fast, though.

But before we look ahead, let's take one more look around at what's happening with VR storytelling. Every Hollywood studio you can imagine—21st Century Fox, Paramount, Warner Bros.— has already invested in virtual reality. They've made VR experiences based on their own movies, like *Interstellar* or *Ghost in the Shell*, and they've invested in other VR companies. Hollywood directors like Doug Liman (*Edge of Tomorrow*) and Robert Stromberg (*Maleficent*) have taken on VR projects.

And the progress is exhilirating. Alejandro González Iñárritu, a four-time Oscar winner for Best Director whose 2014 movie *Birdman* won Best Picture, received a Special Achievement Academy Award in 2017 for a VR short he made. Yet, *Carne y Arena*, which puts viewers inside a harrowing journey from Mexico to the United States, is nothing like a movie, or even a video game. When it premiered at the Cannes Film Festival in early 2017, it was housed in an airplane hangar; viewers were ushered, barefoot, into a room with a sand-covered floor, where they could watch and interact with other people trying to make it over the border. Arrests, detention centers, dehydration—the extremity of the human condition, happening all around you. In the announcement, the Academy of Motion Picture Arts and Sciences called the piece "deeply emotional and physically immersive." (The last time a movie received a similar special-achievement award from the Academy? That'd be 1996, for a little something called *Toy Story*.)

With work like that, it's little wonder that film schools are filled with aspiring writers and directors who don't want to limit themselves to "film," but instead want to push their creativity to the edge in a narrative environment where there *is* no edge. VR story-telling's pipeline may have begun with a few documentarians and animation veterans, but it's growing every day.

Okay. So. A few years ago, I was interviewing then–Oculus CEO Brendan Iribe. (Being owned by Facebook, Oculus no longer has a CEO, so Iribe is now heading a major division in the company.) By that point he and I had spent hours together as part of my reporting on the company for a magazine story. Iribe tended to get so excited about VR stuff that Oculus was working on that he'd *really* want to talk about it—but he couldn't talk about it candidly with me. Instead of asking to go off the record so that he could give me details, though, he had a habit of communicating his excitement in more coded ways. And this time, he did with a movie. "I just watched *The LEGO Movie* in 3-D," he said. "I went with my girlfriend's six-year-old. And I'm like, imagine when he's looking at the movie and it's right here"—he motioned around his face—"and he's able to look around and get close. And when you get close, the little Lego guy looks and turns and says, 'Hey back up a little bit. We have a movie going on here, back up.' He actually looks you right in the eye. Kids are going to think the Lego is real."

We've seen part of that, thanks to *Henry*. And we know a little bit about the power of eye contact. But what about when everything else comes together? It's starting to. In *Asteroids!*, Baobab's follow-up to *Invasion!*, you play a service robot on an alien ship; there's also a doglike robotic creature there, who gives you a ball so that you'll play fetch with it—which you do via your hand controllers, your robotic tentacles snaking out as you toss the ball around the ship. It's a small moment, but an effective one, combining as it does embodiment and direct interaction with a character.

Does it change everything we thought we knew about film?

Not yet, but it's less frivolous than it seems.The future of VR storytelling is going to leverage our own ability, our own *willingness,* to believe that we're someone else. That changes the nature of fantasy and escapism in some fundamental ways. Reading a book, we naturally imagine ourselves in the protagonist's role; in movies, that gets more difficult because we're looking at them. But to truly become a character—to see through their eyes, to take on their properties, to exist within their world—that's an out-of-body experience like nothing else we've been able to accomplish without the use of some powerful psychedelics. Like those psychedelics, though, walking in those shoes and sharing those emotional connections promise to change the way we live our lives, long after the VR experience is over.

That's not happening now because, well, baby steps. Storytellers don't want VR experiences to be too flexible while they're still learning the rudiments of the form. "If you give people too much ability to interact with things, it's often harder for us to tell a story," Penrose founder Eugene Chung said when *Allumette* was first unveiled. As hand controllers get lighter and as obstacles to hand and social presence start to fall away, creators will be devising new ways to take advantage of our newfound freedom. The visual language of VR storytelling won't evolve slowly the way that of film did; it will do so at light speed.

And if we're already experiencing tenderness and joy and grief just by virtue of being in the same space watching other characters, imagine how much more powerful those emotions will be when we start interacting more with characters.

Or with real people—but that's an idea for another chapter.

# 4

# EMPATHY VS. INTIMACY
## WHY GOOD STORIES NEED
## SOMEONE ELSE

Every spring, a thousand curious people descend on Vancouver, British Columbia, to watch what might be the world's most expensive live-staged PowerPoint presentations. The TED Conference is the annual flagship event of its nonprofit namesake, TED (which stands for Technology, Entertainment, Design). The short description is that it's a celebration of ideas; the slightly longer one is that it's a five-day parade of scientists, authors, ex-presidents, and other "thought leaders," as well as Bono, delivering short speeches and lectures to a rapt audience of people who had a spare eighty-five hundred dollars to spend on tickets.

"TED Talks" are known for being counterintuitive and charming. They also happen to be internet gold: thousands of TED Talks stream online, and they've been watched well more than a billion times—not as many as a Justin Bieber video (well, *some*

Justin Bieber videos), but at least enough to make you feel like civ-ilization isn't completely unsalvageable. In the thirty-four years since TED began, it's spawned an empire of mini-TEDs in every city and region you can think of. Admittedly, the so-called TEDx conferences aren't always as rigorous as the original; a few years ago, someone in Portland gave a talk called "How to Use a Paper Towel," which, like, *come on*. The flagship remains a hot ticket, though, and every year it gives rise to a handful of TED Talks that legitimately feel like brain food. In 2015, one of those talks came from a guy named Chris Milk.

If you're a music fan, you might be familiar with Milk's work—he's directed videos from Kanye West, Johnny Cash, Beck, and a bunch of other big names. He's also always enjoyed playing with technology and has created a number of experimental interactive projects, like the time he made a video for the Arcade Fire song "The Wilderness Downtown" that featured your childhood home. Yes, *your* childhood home; the video played in a web browser, and when you entered your address, it used satellite imagery and Goo-gle Street View to personalize the video for each viewer. With a creative streak like that, it's little surprise that Milk was one of the first filmmakers to start working in virtual reality. In 2014, he filmed a 360-degree video of the Millions March NYC against police brutality, making it an early example of VR journalism.

By the time Milk got onstage for his TED Talk in 2015, he'd taken VR pieces to the Sundance Film Festival, including a short docu-mentary he'd made for the United Nations that chronicled the life of a young Syrian girl in a refugee camp. His talk was called "How Virtual Reality Can Create the Ultimate Empathy Machine," and it drew on that documentary, as well as the "Wilderness Down-town" video, to evangelize about the power of VR presence. "It's not a video game peripheral," he said toward the end of his ten-minute talk. "It connects humans to other humans in a profound way that I've never seen before in any other form of media. And it can change people's perception of each other . . . virtual reality has the potential to actually change the world."

Obviously, I think he's right, or I wouldn't have written this book. However—and obviously I also think there's a "however," or I wouldn't have written this book—I also think that his TED Talk, as grand as it was, stopped short in a very important way. Empathy is a marvelous quality, and the ability to truly imagine another person's life would make most of us better to be around in general. It's a crucial ingredient of the recipe for human connection. It's the ingredient that matters most for filmmaking, certainly. But it's not the only one. The other one is just as transformative, and possibly even more immersive. It's empathy's fraternal twin, intimacy.

## EMPATHY VS. INTIMACY: APPRECIATION VS. EMOTION

Both of these words are fuzzy, to say the least. Both have decades of study behind them, but both have also appeared on more magazine covers than just about any word, other than possibly "abs" and "Oprah." What they truly mean often says as much about the person using them as the words themselves, so let's try to boil them down a bit so we're starting with a shared definition. To that end, let's see what the *Oxford Dictionary of Sociology* says about both terms.

**Empathy:** The ability to identify with and understand others, particularly on an emotional level. It involves imagining yourself in the place of another and, therefore, appreciating how they feel.

**Intimacy:** A complex sphere of "inmost" relationships with self and others that are not usually minor or incidental (though they may be transitory) and which usually touch the personal world very deeply. They are our closest relationships with friends, family, children, lovers, but they are also the deep and important experiences we have with self (which are never exactly solitary): our feelings, our bodies, our emotions, our identities.

Immediately, you can see a few distinctions. Empathy necessarily needs to involve other people; intimacy doesn't. Empathy involves emotional *understanding*; intimacy involves emotion itself. Empathy, at its base, is an act of getting outside yourself: you're projecting yourself into someone else's experience, which means that in some ways you're leaving your own experience behind, other than as a reference point. Intimacy, on the other hand, is at its base an act of feeling: you might be connecting with someone or something else, but you're doing so on the basis of the emotions *you* (or both of you) feel.

In fact, you can already see that a lot of the VR we've examined up to this point doesn't necessarily induce empathy, but it still falls squarely into the realm of "intimacy." While meditation, visualization, and even (depending on how much of a monster you are) Henry the hedgehog's birthday party may not push you to appreciate what others feel, they do trigger experiences—sometimes articulable, sometimes primal—that touch your personal world. And one type of VR experience perfectly illustrates the surprising gap between empathy and intimacy: live-action VR.

Unlike CGI-based storytelling, which falls somewhere in between game and movie, live-action VR feels much more like the conventional video forms that we're used to from television and movies. Like those media, people have been using VR to shoot everything from narrative fiction to documentary to sports; however, like everything else in VR, it places you at the center of a 360-degree sphere of action. Using it, I've experienced empathy without intimacy, and even intimacy without empathy. (I acknowledge that the last sentence makes me sound like a sociopath; it's not that I *couldn't* muster up empathy, only that some experiences might not actively induce empathy. And at the risk of this parenthetical aside going on far too long, let me just point out that we'll explore this more and more as the book goes on. By "this," I don't mean long parenthetical asides—though, actually, probably those too. At this point, it's just more to see what my editor lets me get away with. Still with me? Great. Let's get back to the rest of the paragraph.) And a quick spin through some seminal examples of early live-action VR can illustrate some of the differences in—and overlap of—empathy and intimacy.

## WE'LL DO IT LIVE

Well before the renaissance of consumer VR, people were using virtual environments not simply for entertainment, but for information. The most prominent of these was Nonny de la Peña, a documentarian and journalist known as "the godmother of VR"; as a senior research fellow and then doctoral student at the University of Southern California, de la Peña devised prototypes and experiences that charted an early course for reported storytelling in VR. In 2007, she used the virtual community Second Life to re-create Guantanamo Bay prison. (In case you don't remember Second Life, it was like the video game *The Sims,* minus the game part. Users could design their own characters, environments, and activities and then interact with other users through a computer interface. It got very popular very quickly, and then got less popular even more quickly. However, the company that maintains it claims that nearly a million people still use it every month.)

At the Sundance Film Festival in 2012, the same year the phrase "Oculus Rift" was uttered for the first time, de la Peña screened a VR experience called *Hunger in Los Angeles.* It used CGI, along with archival audio, to reenact a heartbreaking episode that happened at a Los Angeles food bank on a hot summer day. In the piece, as in real life, an overlong line leads to delays, and a man collapses into a diabetic coma. "Okay, he's having a seizure," a volunteer says as the man spasms helplessly on the sidewalk; a bystander calls 911. The entire time, you're able to walk throughout the crowd of people, seeing their faces, watching their struggle. It's not light fare. This was many people's first experience with virtual reality—Chris Milk himself among them—and the first glimpse at presence proved to be overwhelming. In one video viewable on YouTube, the actor Gina Rodriguez (*Jane the Virgin*) actively weeps while a volunteer helps her take her headset off.

As emotionally affecting as these experiences were, though, they were still clumsy; the computer-generated graphics were simplistic by today's standards. It wouldn't be until early 2014 that live-action video managed to capture the three-dimensional 360-degree visuals that are virtual reality's hallmark effect. A short documentary called *Zero Point* was ostensibly about VR, but it was shot *for* VR, using a ring of ultra-high-definition cameras pointed outward—so that the camera functioned as the viewer's eye, able to see anywhere inside the 360-degree sphere. (The 3-D effect arose from using two lenses for each view, allowing for the same fool-your-brain trick that's been around since the stereo-scopes of the nineteenth century.)

At nearly the same time, Chris Milk was developing his own VR filmmaking pipeline at a new company he'd founded, and that company—now called Within—came into its own with a documentary he created with the United Nations. *Clouds Over Sidra* takes viewers inside the life of a twelve-year-old girl living in Zaatari, a Syrian refugee camp in Jordan. When the experience begins, you're in the desert, footsteps and tire tracks traversing the sand. "We walked for days crossing the desert to Jordan," a woman says in a voice-over, translating young Sidra's words. Over the next seven minutes, you glimpse daily life in Zaatari through a series of vignettes, narrated by Sidra's words: you see her family bustling inside their small house, boys playing first-person shooters inside a small gaming café. As the film continues, the sense of normalcy mounts; whether you're watching men stacking fresh hot flatbread inside a bakery or a cluster of young girls playing soccer, you find yourself lulled by the seemingly universal daily routines. ("Here in Zaatari, unlike home, girls can play football too," Sidra says. "That makes us happy.") Lulled, that is, until a shot fades in and you see Sidra and her family eating a meal—under the shelter of a UN Refugee Agency tent. "My teacher says the clouds moving over us also came here from Syria," Sidra's narration says. "Someday, the clouds and me are going to turn around . . . and go back home."

As a conventional documentary, *Clouds Over Sidra* might seem simple. The brief, static takes feel more like the type of short film you'd see at a museum, in one of those small bench-filled screening rooms you duck into to rest your legs. Yet, as a VR experience, it's anything but simple. The children aren't smiling and gesturing at a camera; they're engaging with you, the visitor to their camp. The clouds rolling overhead—over *your* head—feel at once ominous and hopeful. You haven't just watched a video postcard that Sidra made; you've spent time in Zaatari with her, seeing what she sees. In other words, you no longer have to imagine the emotions of a young girl living in a refugee camp. You've been there with her. And now, to paraphrase our working definition of empathy, you can appreciate how she feels.

## THE MOMENT OF THE MOMENT

About a year before Chris Milk's 2015 TED Talk—before even *Clouds Over Sidra*—I got my first firsthand taste of VR's potential beyond both entertainment and empathy. Oddly enough, it was at yet another annual conference known by its initials: SXSW. The South by Southwest festival in Austin, Texas, has over time become home to three overlapping but distinct conferences dedicated to interactive technology, film, and music. And in 2014, SXSW saw its first influx of VR. In order to promote the upcoming season of *Game of Thrones*, HBO had brought along a VR installation that allowed people to step into what was essentially a vibrating phone booth—but once they put on a headset, everything they saw, heard, and felt made them think they were ascending the seven-hundred-foot-tall Wall from *Game of Thrones*. (During her time in the booth, a colleague of mine yelped and flung herself backward, falling out of the booth and into the arms of a quick-reflexed HBO rep.)

Not far from that, in a converted loft, I met Félix Lajeunesse and Paul Raphaël, a couple of French-Canadian commercial directors who had recently jumped into developing movies for VR. They sat me down, put an early version of a Rift on my head and headphones over my ears, and played for me their first project, *Strangers*.

This wasn't a computer-generated environment, like every other VR demonstration I'd seen up until that point. It was video. For the first time, I was inside a movie. I was sitting in an apartment, the floor littered with instruments and recording equipment; in front of me, a man at a piano smoked a cigarette. It was quiet in the apartment, other than the man idly plunking his way through some half-formed riffs, but it was also overwhelmingly tranquil. Soon, the man started to play in earnest, and to sing. He seemed to not mind my being there, or even to register my presence, so I didn't think it would be rude to look around while he played. I couldn't move—Lajeunesse and Raphaël had captured the scene by placing multiple cameras on a stationary rig—but I could spin around to see the rest of his apartment. Behind me, on the hardwood floor, a dog dozed in the glow of a nearby lamp. "Are there strangers who live in your head?" the man sang. "Are there strangers who walk in your heart?" I somehow felt both invisible and seen. I felt like there was nowhere else in the world I needed to be. And above all, I felt a sense of warmth like I'd never experienced outside of real life. I was with a stranger, and I felt like I'd just gotten a hug from someone I genuinely cared about.

When the short film ended, I took off the headset and looked at Lajeunesse and Raphaël. "This is a weird word to use," I said, "but it felt . . . intimate."

"That's how we think about it," Lajeunesse said. "We think about creating personal experiences for people to live an experience of presence. We don't think of it as 'he's doing a performance'; we think about it as 'he's just with you.' It really is a moment with that person before being anything else."

*A moment with that person.* Think about that for a second. When's the last time you told someone a story about a terrible thing that had happened to you and you used the word "moment"? Moments are small, good things. They're ephemeral, but have lasting effect. To be warm and fuzzy about it, they're tiny slices of human connection. You can interact with dozens of people in a day, from co-workers to store clerks to people on the bus, but the interactions that stay with you are the moments. They're the bits of shared laughter or surprising sincerity, the displays of compassion. Moments are the building blocks of intimacy among people—even people you don't yet have an intimate relationship with. And with VR, those moments can arise when the person you're sharing it with isn't actually there.

## BRINGING FOCUS BACK TO FICTION

Let's back up a second here. Every single story has only one goal at its base: to make you care. This holds true whether it's a tale told around a campfire at night, one related through a sequence of panels in a comic book, or the dialogue-heavy narrative of a television show. The story might be trying to make you laugh, or to scare you, or to make you feel sad or happy on behalf of one of the characters, but those are all just forms of caring, right? Your emotional investment—the fact that what happens in this tale *matters* to you—is the fundamental aim of the storyteller.

Storytelling, then, has evolved to find ways to draw you out of yourself, to make you forget that what you're hearing or seeing or reading isn't real. It's only at that point, after all, that our natural capacity for empathy can kick in. Yet, innovation in this area has been surprisingly rare; once we settled on the idea that simulations happen inside a frame, all we could really do is bring the frame closer to us: bigger screens, louder speakers, the gimmicky hints of immersion that swept movie theaters during the 1950s and 1960s—rumble seats and Smell-O-Vision, misters and hypnotists.

Meanwhile, technology continues to evolve to detach us from those stories. For one, the frame itself continues to get smaller. From news to television to literature, more and more of our media consumption is happening on our phones—or at least something reasonably phone-size, like an e-reader. But the size of the window is only part of it. By nature of its portability, the phone isn't something you lose yourself in, but is a supplement to everything else. It's increasingly common to watch a television show on a phone or tablet, while in line at the bank or on the bus commuting to work. (And when we do use a traditional television, we're often on our phones at the same time, scrolling through social media or news, or even YouTube.) Stories are fighting for our attention, and even when they get it, we've got one cerebral foot out the door.

Stranger still, this distraction has happened while stories continue to become more and more complex. Narratively, at least, stories are more intricate than they've ever been. Long-arc prestige TV series pull scores of characters and numerous storylines along multiple seasons; cinematic universes encompass a decade's worth of movies; fantasy epics can span a dozen books or more. Even sitcoms, once a simple twenty-two-minute dose of laughter, can rival dramas for structural intricacy, challenging viewers with knotted plots and multiple points of view. And all the while, the things we want to consume but haven't yet continue to pile up. There's more than we can ever read or watch, and we want to read or watch it all—but when we do, we binge, often with another frame close by.

Now, with VR storytelling, the distracting power of multiple screens has met its match. There's no Twitter notification or calendar alert or lingering text conversation that follows us into an experience like *Clouds Over Sidra* or Lajeunesse and Raphaël's *Strangers*. There's just us, in there, with other people's stories, giving them our full attention. Given what a rarity undivided attention is these days, that's a miracle in itself. And by experiencing those stories inside the frame, we're left with, as Chris Milk said in his TED Talk, an unparalleled capacity for empathy.

That's not to say that VR has transformed us into a nation of altruists. In fact, Paul Bloom, a psychology professor at Yale who in 2016 ruffled feathers with his book *Against Empathy,* has argued that VR at best offers a pale knockoff of empathy: "The problem is that these experiences aren't fundamentally about the immediate physical environments," he wrote in *The Atlantic.* "The awfulness of the refugee experience isn't about the sights and sounds of a refugee camp; it has more to do with the fear and anxiety of having to escape your country and relocate yourself in a strange land."

He's partly right; despite the 360-degree fullness of what the viewer experiences, VR documentary experiences can offer only a slice of a subject's experience. However, there's evidence that even now, VR's immediacy and immersion do actually lead to increased philanthropy. When the UN screened *Clouds Over Sidra* for potential donors before a humanitarian conference in Kuwait, the organization raised nearly $4 billion—twice what was expected.

The ability to understand another person's experience is clearly a magical ingredient for documentary and narrative stories. The characters you meet take on new depth; it's only natural to become more invested in people's journeys and relationships when you're present in their environments. When a story tries to scare or charm or arouse you, it's all the easier when you're *there*.

But if those stories are simply about other people, then empathy is effectively the experiential ceiling. You can understand a person's life, but you don't share it. When you're in Zaatari watching Sidra's family eat, for example, you see them having a moment together, but you're not included in that moment. You witness it, rather than being part of it. And intimacy—at least once it extends outside yourself—is by nature a shared phenomenon. So the ability to wring intimacy out of VR is going to depend on experiences that involve you, experiences that establish moments.

Just how VR can and will establish those moments is still a matter of exploration and experimentation. Félix Lajeunesse and Paul Raphaël's *Strangers* doesn't impose a story on the experience, nor does it enlist you as a character; depending on how much you assume that the pianist is singing to you, it doesn't even exactly acknowledge your presence. Instead, by inviting you in to another person's solitude, it lets you share that experience in a sustained way. There's a *with*-ness to it, which is what makes it a moment.

That's just the beginning, of course. The first few years of live-action video have been hamstrung by a severe technical constraint—one that we're just now starting to get past. And as we do, we will be able to unlock a new level of intimacy in storytelling.

## THE FULLNESS OF YOU

For all the grandeur that we attach to the name, Silicon Valley is not a grand place. It's certainly not a valley, at least not anymore; while the sobriquet once referred to San Jose and lands south, where chip manufacturers cluster, Silicon Valley has crept northward to colonize the exurban sprawl of the San Francisco Peninsula. Now, tech culture crams itself alongside residential communities and commercial arteries on a narrow swath of land that's bounded by the San Francisco Bay on one side and winding hills on the other. A handful of towns—Menlo Park, Cupertino, Palo Alto—host a matching handful of billion-dollar behemoths.

In Mountain View that juggernaut is Google, but the city is also home to dozens of other companies, tucked into small offices along quiet, curving roads. Inside one of those, that of an imaging company called Lytro, CEO Jason Rosenthal hands me a headset. "You'll hear about eight seconds of audio before you see anything," Rosenthal says, "then the lights will come up."

When the lights come up, I'm watching an astronaut step off the ladder of his lunar lander, down onto the surface of the moon. The earth hovers over his shoulder in the jet-black sky. "That's one small step for man," he says, "one giant step for mankind." *One giant step?* I think. *I thought it was lea*—and as if on cue, a voice behind me barks "CUT!"

But then even more lights come up . . . and I realize I'm standing on a film set, watching conspiracy theorists' favorite hobby horse come to life: Stanley Kubrick directing Neil Armstrong's "fake" moon landing. The astronaut and lander are still there, but the jet-black sky overhead is gone, replaced by girders and rigging and the other trappings of a Hollywood soundstage.

It's a short video, just enough to sell the joke, but the joke isn't the real point. The point is that this is a 360-degree video unlike any other. When I lean to the side to get a better look at the lunar lander, my perspective changes; when I crouch down to get out of Kubrick's way, he looms higher in my field of vision. This is a video, but I'm moving within it more like it's real life. This is "lightfield" technology, and it's one of a handful of ways companies are trying to game VR video's biggest limitation.

Remember this? That's the diagram of the six different axes on which your head can move, the so-called six degrees of freedom.

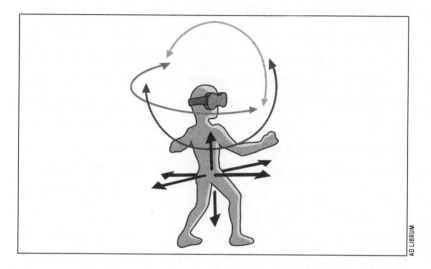

AD LIBRUM

The curved lines are all the ways your head can rotate—tilting, nodding, turning—and the straight lines are how it can change position within space. Physicists refer to these as "rotational" and "translational" movement; the VR industry prefers "rotational" and "positional." (Personally, I prefer "rotational"/"locational," because it rhymes so nicely, but I also lack a degree in physics, so you might want to side with the scientists on this one.)

As I mentioned in Chapter 1, high-end VR headsets are able to track both rotational and positional motion, but mobile headsets like the Samsung Gear VR and Google Daydream View can handle only rotational tracking. That's because they rely on a smartphone's internal accelerometer and gyrometer. If you load up *Henry* on your Oculus Rift, you can peek underneath Henry's living-room table; not so if you watch it on your Gear VR. Because of that limitation, mobile VR tends to be a seated and stationary activity, while PC-driven headsets allow so-called room-scale VR in which you can roam around on foot IRL in order to explore a virtual space.

However, we're on the cusp of a big change here as well. In 2018, all-in-one "standalone" headsets are coming to market that are able to track your position in space, meaning they allow all six degrees of freedom. (The jargony phrase is "6DOF tracking," which you can call it if you're hanging out with VR enthusiasts. And yes, it's pronounced "doff.") Those may well end up becoming the future of everyday VR: They contain dedicated displays and computer processors so that you get all the 6DOF-tracking goodness of a PC-powered headset, without the actual PC—or the cables that tether you to it.

Better VR at lower prices is great news for everyone. Yet, just because wireless headsets can be tracked positionally won't change the fact that live-video VR still can't take advantage of it. That's always been a limitation of any photograph: you might be able to use two almost-identical images to trick your brain into seeing depth, but that depth isn't real. The same thing holds true for a 360-degree photo or video. *Clouds Over Sidra* may induce a 3-D effect, yet in reality it's about as deep as a 3-D movie—you can't change your perspective within the spherical frame, or change your location within it. Wherever the camera rig was placed, that's your vantage point, for better or for worse.

Lytro's moon-landing video, though, utilizes an entirely different kind of camera rig. Rather than arraying a bunch of outward-facing cameras around an axis that's meant to be the viewer's perspective, Lytro's Immerge camera is a big flat hexagonal panel with sixty small cylindrical camera lenses embedded on it—like a fly's eye, kind of. Here's the fun part: each of those lenses is able to capture not just light, but the direction in which light is traveling. By shooting the same scene from multiple angles, the camera is able to record the scene in a way that knows exactly what you'd see from any point within that scene. Software then stitches it all together into a VR video that allows you to examine your surroundings from multiple angles. (It's much much much *much* more complicated than I'm making it sound. Lytro's founder wrote a doctoral dissertation on the topic of lightfield capture, and let's just say that it's beyond most of us. Also, don't get your hopes up about picking one up for a quick shoot; the camera system in question is six-figures expensive.)

The moon landing was an early demo, but Lytro is already seeing the fruits of its work. Chris Milk's company, Within, used the system to create *Hallelujah,* a music video in which a singer performs Leonard Cohen's stirring song. The effect is extraordinary; while the singer doesn't follow you with his eyes the way Henry the hedgehog does, the sense of *with*-ness is strong—and when the lights come up, as they did in the moon-landing demo, and you see for the first time the choir behind the singer and the splendor of the church you're standing in, that with-ness is coupled with straight-up awe.

Lytro's lightfield method is just one way people are creating "volumetric" video—video that isn't a flat plane, but a full space. Some companies are recording human performances on a green screen and then digitizing those performances and placing them into other video environments that have *also* been digitally re-created, allowing for movement within the environment. (Imagine motion-capture performances like Gollum in The Lord of the Rings, just without the need for a mo-cap suit.) Facebook has even developed two small cameras that promise video you can move within, for more indie-film-friendly prices. VR video will soon feel far freer, and more comfortable, than it does now.

As welcome as the technical triumph is, though, the true impact may be what it means for storytelling's ability to establish intimacy. Merely sitting in a musician's studio, as I did in Félix Lajeunesse and Paul Raphaël's *Strangers*, was enough to create a moment. But what if I had been able to get up, walk across the room, and sit on the floor with the napping dog?

We're starting to get answers to that question, as more and more volumetric VR video experiences are screening at film festivals. In one, you accompany a Holocaust survivor to the Polish concentration camp where he spent his childhood; standing with him in the barracks and crematorium, looking at the empty beds, is more personal and affecting than any documentary or tear-jerking feature film. In another, created by Nonny de la Peña's studio Emblematic Group in collaboration with PBS series *Frontline,* you stand with an ex-convict in a solitary confinement cell. Despite some moments where you can tell that he's been digitized—the edges of his clothes flicker at times—leaning toward him while he talks about the long-term effects of sensory deprivation is chilling. It's sympathy, empathy, and intimacy, all wrapped up in one VR package. Perhaps most exciting, each of these experiences comes from different studios, all of which are using different methods to deliver new depth to video—and new depth to the connections we make within those videos.

## EXPERIENCING OUR LIVES—TOGETHER

What video still can't do, though, is bring more people together inside VR, the way Ray McClure's singing-multicolored-blobs-at-each-other tag-team project VVVR does. That's why even VR filmmaking powerhouses like Within are moving beyond mere documentary and narrative and trying to turn storytelling into a shared experience.

Make no mistake: storytelling has always been a shared experience. But I don't mean sitting in the dark with friends listening to a creepy tale or watching a movie together. I mean being conscripted into the story, or even *being* the story. And that's exactly what I'm looking for when I walk into the Within office on a sunny afternoon in May.

Like so many of the companies in Los Angeles's burgeoning VR scene, Within is based in Culver City, a hybrid of hip and Hollywood that's also home to more conventional media powerhouses; Sony Pictures's massive studio lot is less than a mile away. Fittingly, Within's director of original content also hails from the world of conventional filmmaking. Jess Engel got her start producing independent movies but landed at Within in 2016 to spearhead the company's narrative efforts. Later in 2017, she'll venture out to launch her own VR production company, but today she's been nice enough to indulge my request to come see the company's newest work.

However, "see" doesn't quite cover it. When I visit Engel at Within, she's just returned from the Tribeca Film Festival, where she helped bring festivalgoers through an interactive VR project called *Life of Us*. And now, we're going to experience the title the way Within intends it—together. Once I have my HTC Vive headset on, and the controllers in my hands, Engel heads into a small office just down the hall, gets her own gear on, and launches us into the experience.

My first reaction is surprise. I've seen just about everything Within has created, and the vast majority of it is video. Now, though, I'm not just a viewer, but a character—and not even a human one at that. I'm a brightly colored, polygonal, almost origami-like amoeba. And so is Engel! "Can you see me?" she asks in my headphones, the software distorting her voice into a burbling high-pitched coo. "There you are!" I say, laughing as my own pitch-shifted voice echoes in my ears, a split second after I say the words. We float around each other, twitching in what I can only assume is a totally amoebic fashion and laughing hysterically.

Then, in a flash, the scene changes. Now we're some sort of primordial marine tadpole things, blowing bubbles and swimming together through the sea while rays of light pierce the ocean's surface above us. We're moving as if on rails, free to move our head and hands—if we had hands, that is. But soon enough, we do: now we're lizards racing two-legged across a desert, a *T. rex* chasing us. And again; we're flying now, fire-breathing pterodactyls soaring together over volcanoes. Once more—this time we're gorillas running in grasslands, besieged by baby monkeys trying to hitch a ride on our galloping simian bodies, swatting the cute pests off each other's shoulders and arms as we race together toward the next evolutionary step.

Then we're *humans*, dark-suited office drones running in a sea of look-alikes through a cityscape, papers flying from our briefcases. *So much for evolution,* I think to myself. As we run, the city starts to become darker, more futuristic looking. I look at Engel's human, and she's wearing a headset; we both are, with digital devices strapped all over ourselves. Everything goes black. The music stops. Our bodies fall apart. And when a light blinks back on, we're both female robots ("The future is female," Engel reminds me later), dancing to a Pharrell Williams song that the artist recorded especially for *Life of Us*. All around us, we're surrounded by the creatures we once were, from amoeba to hyper-technologized posthumans—until finally, everything goes black again, and big block letters appear in front of me: WAKE UP.

Like so many VR experiences, *Life of Us* defies many of the ways we describe a story to each other. For one, it feels at once shorter and longer than its actual seven-minute runtime; although it seems to be over in a flash, that flash contains so many details that in retrospect it's as full and vivid as a two-hour movie. There's an almost paradoxical compression to the way I think about it now— one that I can only compare, oddly enough, to the one time I went skydiving.

When I jumped out of that plane on a clear July morning, I was in free fall for less than two minutes. Yet, that short time has stayed with me in stunning clarity for years—not as a fluid sequence, but as a haphazard collection of micromemories, each one a gimlet snapshot: my feet flailing, the wind buffeting my cheeks, the plane shrinking away behind me. It's not a holistic experience anymore; it's a slideshow. *Life of Us* has stuck with me much in the same way: the volcanoes, the laughter, Jess–gorilla's face looming near mine as she plucks a monkey from my arm.

There's another thing, though, that sets *Life of Us* apart from so many other stories—it's the fact that not only was I in the story, but someone else was in there with me. And that someone wasn't a filmed character talking to a camera that I somehow embodied, or a video game creature that was programmed to look in "my" direction, but a real person—a person who saw what I saw, a person who was present for each of those moments and who now is inextricably part of my odd, shard-like memory of them.

I know Engel has gone through this journey dozens, maybe hundreds of times, so when we sit down to talk afterward, I ask her the question that I can't stop thinking about: Is VR—and *Life of Us* in particular—still something that she can lose herself in?

"It's the moments when so many of my senses are engaged," she says. "My voice. My eyes. My body. Even though I've done it, it's still different, because we're having a different experience together. It's like a restaurant. You can go to the same restaurant over and over again. But every time you go to that restaurant it's going to be different because the person you go with is different— even if you order the same thing, it's going to be different. That's what shared experiences can do."

"What's interesting about this tool," Engel says, pointing to a headset, "is that it has no meaning. It's just some hardware. But the way you use it has a lot of meaning."

"It's the restaurant," I say. "The restaurant is the headset."

She smiles. "You go there, and it's about who you're with and how you engage with each other."

Who you're with. How you engage with each other. For the first time in our lives, we're faced with a technology that actually puts us somewhere else. Physically, mentally, and, yes, emotionally. It doesn't merely demand our attention, or capture our imagination, or make us think about it afterward; it does away with sentences like *I read a book about . . .* or *I played a game in which . . .* and replaces them with *I evolved from a microbe into a shimmering posthuman light.* And best of all, it lets us experience that story with someone else—and, in doing so, find a new kind of intimacy with them.

# 5

# WHAT TO DO AND
# WHO TO DO IT WITH

## HOW SOCIAL VR IS REINVENTING
## EVERYTHING FROM GAME NIGHT
## TO ONLINE HARASSMENT

N OW SEEMS as good a time as any to make a confession: My whole life, I've loved the idea of role-playing games like Dungeons & Dragons. I was infatuated by the dice, the character sheets, the intricate rules. When I was in elementary school, I'd buy pads of graph paper just so I could draw dungeons, plotting out labyrinthine corridors and catacombs that would ensnare my imaginary adventurers. I'd create characters, inventing elaborate backstories that explained why a gnome fighter would insist on trying to wield a sword that was nearly as tall as he was. I'd pore over books like the famous *Monster Manual*, reading about how many hit points a Gelatinous Cube had and memorizing the various types of illusion the carnivorous Rakshasa could cast. (One illusion I don't have, clearly, is how cringe-worthy that last sentence is.)

Yet, I never played them.

Although I loved the *idea* of the game, the actual experience left something to be desired. Not enough of my friends ever wanted to play, and when we managed the rare quorum, it generally fell apart within an hour, whether because of boredom, confusion, or the diuretic power of Mountain Dew. I tried going to game stores but always felt self-conscious about joining a group of more experienced players. And sure, yeah, there might have been some shame in there as well—but regardless of the cause, the upshot was the same. I never broke through the beginning stage, never knew the thrill of banding together with my merry adventurers to defeat evil and lay claim to treasures untold.

Now, though, I'm standing in a tavern, trying to do just that. Well, it's a *virtual* tavern, situated within Altspace, one of the early multiuser social apps available in VR. Five of us are gathered around a giant table: me, three other D&D players, and Tim, a "dungeon master" who's taking us through the game. I'd say we're a motley crew, but that's not quite the case. Altspace offers only a rough handful of avatars you can select to represent yourself, and three of the five of us chose to be robots of various colors. The other two—me and a guy named Zyan—look as expressionless as the robots, thanks to the unchanging expressions on our avatars' faces.

Tim asks us to explain a bit about our characters, and the adventure picks up where it left off last week. (Fair warning: if the rest of this paragraph sounds like *Lord of the Rings* fan fiction, that's because that's basically exactly what D&D is.) I'm playing a human druid named Cale Firborn; my fellow glory-seekers and I have taken on jobs as hired labor for a caravan of traders. We've been traveling by wagon for the past three weeks, working by day and by night trying to figure out who among us might be members of the Cult of the Dragon. Evil's afoot in the land, and we're hot on its trail. We've finally arrived at an inn just as rain starts to pour down, only to be turned away by the innkeeper—and laughed at by the group of aristocrats sitting in the inn's common room.

We've got a hunch they're holding the innkeeper hostage. Besides, sleeping outside isn't an option; it's freezing, and our horses might not survive. "We've got to get in there," says Dust, a gray robot sitting to my left who's playing an elf fighter.

As a druid, I have the power to turn into a brown bear, so I tell Tim that I'm going to try to break down the door of the inn with my ursine might and chase the haughty aristocrats out of the inn—at which point my nonbear friends will rush in to save the day, and the grateful innkeeper will surely extend us some much-needed hospitality. "Sounds good," he says. "You're going to need to roll a strength check."

Every game of Dungeons & Dragons, as eleven-year-old me would have been only too glad to tell you, revolves around dice. Every time you want to do something—listen for a rustle in the woods, jump across a chasm, cast a spell—you roll dice. Some of them are the usual six-sided dice you'd see in any Monopoly or craps game, but others are more exotically shaped: four-sided, eight-sided, ten-sided, twelve-sided, and twenty-sided dice all make frequent appearances at a role-playing game table. There's even a 120-sided die, known to mathematicians as a disdyakis triacontahedron, though it doesn't have many uses for owners beyond maybe determining what age you'll finally lose your virginity.

In this case, a "strength check" requires me to roll a twenty-sided die: the result will determine whether the door can survive intact, or whether the inn becomes a bear sanctuary. I look around the table. No dice. I look over at Tim, and he silently points up. I look up, and I see why: dice of every possible shape are hovering over the table. I press a button on the Oculus Touch controller in my hand and select the virtual icosahedron. It falls to the table and tumbles freely, finally stopping on a fourteen. That's one shy of what I need to break through the door, but I'm not disappointed. I've finally found the dream: pickup game night from the comfort of a headset.

## WHAT YOU DO— OR WHERE YOU DO IT?

If you're going to work on the future, it helps if you look like you're from the future, and Eric Romo, the cofounder and CEO of Altspace, certainly does. It's not because of his clothes, which tend toward business casual, and it's only partially because he's tall and lean. It's mostly because he's totally hairless. If you have to bet on someone to figure out what hanging out will look like in 2023, go with whoever's most aerodynamic.

VR isn't even Romo's first bet on the future. When he was finishing up his master's degree in mechanical engineering, a professor emailed him on behalf of two men who were recruiting for a rocket company they were starting. One of those men was Elon Musk, which is how Romo became the thirteenth employee at SpaceX. Eventually, he started a company focusing on solar energy, but when the bottom fell out of the industry, he shut down the company and looked for his next opportunity.

This was the end of 2012—exactly the time when VR was popping back up on people's radars. Romo spent the next year and a half researching the technology and thinking about what kind of company might make sense in a new VR-enabled world. He had read *Snow Crash*, like everyone else in the VR world, but he also knew that our hopes for a VR future could very well end up like the famed flying car: defined—and limited—by an expectation that might not match perfectly with what we actually want. "The biggest piece of baggage people take from those books is that VR is a place that you *go*, and how fantastical it is," he says. "'Wow, it's so cool that I can look at this club that is a big, black sphere that I can walk into!'"

VR, Romo and his cofounders think, isn't about walking into that big black sphere. The space you're in won't matter nearly as much as who you're there with. So Altspace doesn't want you to come in, explore everything, and then leave; it wants you to come back because you liked what you did. When you connect to Altspace, there's a relatively limited number of environments you can be in, and most of them are decidedly nonfantastical: a tavern, a nondescript nightclub, a modernist house, a home theater, a meditation dojo. These aren't destinations but conduits, places that allow you to host a variety of activities, from D&D games to a screening of *Lawnmower Man* (which, as you can probably guess, has happened more than once).

Similarly, your avatar—the character you appear as—is surprisingly simple. Even the most middle-of-the-road video game these days allows you to tweak your character to look exactly how you want it to, from jawline to eye color to brow shape to facial hair to body type. In Altspace, you choose only three things: which one of six avatars you want to be, its skin tone (or, if a robot, its metal tone), and the color scheme of its outfit. That's because customization is, in Romo's view, a "time vacuum." He'd rather spend time figuring out how to let you know about the things to do, and to find people to do them with.

The first part of that is at least imaginable: when you launch Altspace, a calendar pops up in front of you listing the featured events happening in the coming days, as well as the public rooms where people are hosting games or chats. The second part, however, feels like the biggest obstacle facing Altspace, and other platforms like it. Romo puts the number of users at the tens of thousands, but that's not always the feel when you visit Altspace; more often than not, you find a handful of gathering spaces, each with a dozen or so people playing a game or just talking around a virtual campfire.

The busiest times, by far, are reserved for what Romo calls Altspace's "marquee events," which have emerged as the platform's most consistent draws. During the 2016 presidential election, the platform partnered with NBC News to host "Virtual Democracy Plaza," an environment where people could gather to watch the Trump-Clinton debates or attend talks by Al Roker or *Meet the Press* host Chuck Todd. Podcasters record episodes in Altspace; improv comedy troupes do shows. And perhaps most enduringly, the musician and comic Reggie Watts held a monthly residency, doing shows in Altspace every few weeks. The company even granted him a custom avatar, instantly recognizable down to its afro, beard, and suspenders. The first time he performed, in May 2016, the company touted that more than one thousand people attended the event, at the time making it the most people ever gathered together in a VR space at one time.

Watts is likely best known as the bandleader on James Corden's late-night CBS talk show, where he asks each guest a single bizarre question. ("When you think about sensuality," he asked Jeff Goldblum, "do you think it's mostly about listening, or about responding in real time to something ineffable?" To Naomi Campbell, he got a bit more prosaic: "If you had to choose between muffins, cookies, crackers, crisps, or just a good time inside of a boat, which one would you choose?") If you think his non sequiturs are strange on broadcast network TV, though, you should see them in his live show, where his improvisation truly thrives. He alternates between unscripted cerebral riffs and on-the-fly music, creating beatbox-driven songs using a keyboard and a looping machine. The resulting blend—part science fiction, part meta-comedy, and part late-night jam session—may confuse, but it always entertains.

Amazingly, the feeling of those shows translates into VR. Some of that is due to some surprising physical realism. In VR, Watts is walking back and forth on a nightclub stage; in real life, he's at home in Los Angeles, wearing a fifteen-hundred-dollar VR motion-capture suit. Seeing his avatar move in Altspace—head nodding, knees bending slightly as he sways side to side—doesn't increase your own sense of presence, but it makes Watts far more *present* than the other user avatars in Altspace, all of whom move about with a slightly unnerving slide, like so many robot butlers.

There's more to it than mere physicality, though. Watts's dance-like-nobody's-watching aesthetic has always felt like the product of a single overclocked brain. There's a connection with the audience, as with any good performer, but because so much of what he does boils down to "puttering around talking to himself and singing," it doesn't take more than a decent approximation of his bodily self for his other elements to come through fully. And happily, his VR act keeps the musical elements: Watts places hand controllers atop his keyboard so that he can find it while wearing a headset (just look for the floating controllers!).

For all their similarities, the real-life and VR versions of a Reggie Watts show exhibit two stark differences, both of which provide hints of what a future of shared VR experiences might entail. Just as with online multiplayer video games, allowing a large number of people to act and interact in an online space is a staggering technical lift for any company. Altspace, however, wants to be able to accommodate hundreds or thousands of people at a single time, so it developed a clever work-around: it subdivides a VR audience into groups of thirty or so, places those groups in identical VR nightclubs, and then essentially clones Watts's onstage VR avatar across all those identical rooms. Your experience isn't like an arena show, or even a small concert hall, but a much cozier venue with a couple of dozen attendees. While you wouldn't want to see Metallica or Beyoncé giving everything they have in a tiny cabaret, it's the perfect environment for certain types of performances—like, say, a musical comic weirdo. Altspace claims it can handle up to forty thousand users this way, using various types of VR environments to best suit the event in question. As for Watts's experience as the cloned avatar, performing in dozens or hundreds of virtual rooms simultaneously, it's not as overwhelming as you might think. He sees the crowd in only a single room.

What *is* overwhelming is an emerging social behavior that first took root in conventional social media, but takes on a new dimension in VR. If you've watched a Facebook Live video, you've likely seen thumbs-up "like" emoji—as well as other Facebook-sanctioned reaction emoji such as hearts and angry faces—float across the video, straight from viewers' thumbs to your smartphone screen. Altspace enables a similar real-time response: users can send smiley faces, hearts, or clapping hands streaming upward from their avatars' heads. If you're one of thirty people in the room, it's the virtual version of holding up a lighter or a smartphone screen: a perfect way to signal your appreciation without hooting into your microphone and disrupting the show.

If you're Reggie Watts, though, you see *all* the emoji reactions. From *all* the people in *all* the rooms—the air thick with hearts and smiles and clapping hands, floating heavenward through a ceiling that doesn't even physically exist.

## SOCIAL VR, PERSONAL SPACE, AND AGGRESSIVE ARCHERS

As long as our computers have been able to talk to each other, we've been using them to do the same thing. For me, that happened my first year at college, when my friend and I learned how to "ntalk" with students at other schools. Of course, this was in 1993, when the "World Wide Web" was a mind-shattering innovation and using the internet was mostly about typing various nonsense words like "telnet." But still, even then, the promise of meeting people—specifically, in our case, meeting women—immediately eclipsed every other reason we'd been using computers. Video games? Writing papers? Those were fine, but they weren't going to help us hook up.

We weren't the only ones. Over the first three decades of the internet's existence, its evolution was marked by the rise and fall of social communication. First came "bulletin board systems" and an enormous network called Usenet (think Reddit without any graphics), then chat rooms. After that, instant messaging—basically the descendant of my beloved "ntalk"—allowed for back-and-forth conversations in real time between two people. Finally, social networks came along and mashed everything together: now platforms like Facebook, Twitter, and Snapchat allow instantaneous communication between and among every kind of social group you can think of.

As our online social tools have grown in speed and scale, though, so has the ability to use those tools to be an asshole. Back in the day, "trolling" just referred to pursuing a provocative argument for kicks. Today, the word is used to describe the actions of anonymous mobs like the one that, for instance, drove actor Leslie Jones off Twitter with an onslaught of racist and sexist abuse. Harassment has become one of the defining characteristics of the internet as we use it today; a few years ago, the Pew Research Center found that 40 percent of internet users—and 70 percent of those eighteen to twenty-four years old—have experienced online harassment to some degree.

Up to now, that harassment has happened at a remove. It's words and pictures that you see on your phone or computer screen. But with the emergence of VR, our social networks have become, quite literally, embodied. What was once an anonymous commenter is now an anonymous avatar—in the form of another being standing right in front of you. What was once a tossed-off insult or slur is now the ability to invade your personal space.

As soon as high-end headsets were released and VR started attracting larger numbers of users, those users began virtually coexisting. Some people found the combination of presence and online anonymity to be an emboldening recipe. In late 2016, a woman named Jordan Belamire published an article on the online

## Skeptic's Corner: Personal Space

**You:** Personal space? Nope, sorry, not buying it.

**Me:** Explain.

**You:** Okay, well, first of all, *it's not you*. Like, you look down and don't even necessarily see legs, so why would it matter how close someone gets?

**Me:** So it turns out that a lot of people have thought about this for a long time. There's even a term for the field of study: proxemics. Most of the big proxemics studies happened decades before VR came along, but in 2001 a group of researchers at the University of California–Santa Barbara decided to test how being in VR affected people's concept of personal space. They designed a simple virtual room and then gave volunteers a headset and asked them to walk up to the other person they saw in the room and read his nametag. The "person" in question, though, was a virtual man, basically a simple video game character.

Like so many other psychological studies, the volunteers thought they were there for one reason (to have their memory tested), but the researchers really wanted to see how much space people would give the virtual man. And to figure that out, they programmed the character to have one of five levels of eye contact: they ranged from its eyes being closed to staying fixed on the volunteer and turning his head to follow the volunteer. Not surprisingly, volunteers gave a wider berth to the virtual man when he looked directly at them and followed them with his eyes. So yes, there's personal space in VR.

    (Male and female volunteers differed in one pretty fundamental way, though. When men approached the virtual man from the front, the distance they maintained from it wasn't based on how the character was looking at them but on how much "social presence" the character had—in other words, how much the volunteers believed the character was actually conscious and aware of them. Men, it turns out, don't maintain eye contact with other men and so simply didn't notice the character's gaze as much as female volunteers did.] ∎

platform Medium detailing her first experience in VR—which was also her first experience with virtual harassment.

Belamire (a pseudonym) had used her brother-in-law's VR system to try an archery game called—I'll wait while you guess this one—*QuiVr*. *QuiVr* works via the magic of hand presence: one handheld controller acts as your bow, the other as your arrow, and you play by miming the stringing and loosing of arrows in order to snipe undead creatures from atop the ramparts of a mountain castle. After enjoying the game's single-player mode, she decided to play the next round in multiplayer mode.

A quick overview, just in case you know nothing about video games: Multiplayer gaming is just what it sounds like, though more than one person playing the same game simultaneously can take many forms. Some classic arcade games of the 1980s allowed people to play together cooperatively; *Gauntlet*, for example, featured four joysticks, encouraging a merry quartet to explore a dungeon and fight against common enemies. Fighting games like *Street Fighter* were designed for one-on-one competition, in which two players engaged in high-flying, fireball-throwing hand-to-hand combat. However, the advent of the internet moved things online, and now most multiplayer games are decentralized networked experiences in which you play against, or alongside, people you never actually see. For a game to become a legitimate cultural phenomenon—think *Halo, Call of Duty, World of Warcraft, League of Legends, Overwatch*—online multiplayer is an absolute prerequisite.

However, online multiplayer gaming is also an incubator for some pretty vile behavior: not just what could be considered "unsportsmanlike" play, but virulently racist and sexist speech; gender harassment that can begin as soon as a player's name or voice indicates the player is female; and even rape threats and other violent overtures. (It's not all that different from the rest of online harassment, though the fact that multiplayer gaming generally includes voice communication makes the harassment significantly more visceral than images alone might be.)

When Belamire began the multiplayer version of *QuiVr*, she was paired with other gamers who looked exactly the way she did in the game: a floating helmet with a quiver floating behind it where the back would be, a hand clutching a bow, and a free hand that could reach back to grab arrows. No face, no hair, no clothes, no body—just the minimal elements necessary to signify an archer. If it weren't for usernames and voices, anonymity would be complete. However, Belamire was using her microphone, and so the other players knew she was a woman. One of them, whom she dubs BigBro442, began acting creepy.

> In between a wave of zombies and demons to shoot down, I was hanging out next to BigBro442, waiting for our next attack. Suddenly, BigBro442's disembodied helmet faced me dead-on. His floating hand approached my body, and he started to virtually rub my chest.
>
> "Stop!" I cried. I must have laughed from the embar-

rassment and the ridiculousness of the situation. Women, after all, are supposed to be cool, and take any form of sexual harassment with a laugh. But I still told him to stop.

This goaded him on, and even when I turned away from him, he chased me around, making grabbing and pinching motions near my chest. Emboldened, he even shoved his hand toward my virtual crotch and began rubbing.

"As VR becomes increasingly real," she wrote, "how do we decide what crosses the line from an annoyance to an actual assault? Eventually we're going to need rules to tame the wild, wild west of VR multi-player. Or is this going to be yet another space that women do not venture into?"

The response to the article was swift, and strong. The two men who had created the game published a long blog post expressing remorse for what Belamire had experienced and also apologizing for their failure to anticipate the episode. When creating the game, they wrote, they had designed a "personal bubble" so that players couldn't wave their hands in another player's face to block their view—the offending hands would simply disappear—but they hadn't thought about extending that bubble to the body. "How could we have overlooked something so obvious?" they wrote. As soon as they read Belamire's article, they immediately updated the game to enlarge the personal bubble.

However, they wrote, there was more to contend with:

We would like to float a possible way of thinking for the VR development community to consider as we grow. It consists of two parts. One, that we should strive to prevent harassment from happening in the first place, of course.

But second, when harassment does happen—and I see no way to prevent it entirely so long as multiplayer experiences exist—we need to also offer the tools to re-empower the player as it happens.

I don't know if we are right in this belief, but it seems a reasonable one to us—if VR has the ability to deprive someone of power, and that feeling can have real psychological harm, then it is also in our ability to help mitigate that by dramatically and demonstrably giving that power back to the player before the experience comes to an end.

*QuiVr* wasn't the only multiplayer VR experience to consider such questions. A few months before, the creators of an app called Bigscreen had announced a measure much like what *QuiVr*'s creators promised: any user who got within a certain radius of another user would simply disappear. And by now, any multiplayer VR experience that wants to attract users—or, in these days when everything is free, investors—is introducing some sort of privacy or anti-troll measures.

Altspace is among them. After a well-publicized instance in which a female journalist wrote about being swarmed and "kissed" during her first time on the platform, Altspace rolled out its own personal space bubble. That joined other options Altspace had made available to users, such as the ability to mute and block other players, as well as making sure that an Altspace representative is available inside VR every minute of the day. "In the early days we were very pleasantly surprised that we didn't have lots of reports of harassment or anything like that," says Eric Romo. "But as soon as the user numbers started to spike, the number of targets started to go up—so that's where you see things like the block and mute and moderation, and the bubble of personal space, which I think have largely started to become really effective. Reports of harassment are really way down, and I don't know if we've had one in a while. So it's been much better, but I think it's something we have to continue to work on."

This is more than a matter of a space bubble, though. The question that faces the developers of *QuiVr*, and Bigscreen and Altspace and every other VR tool that brings people together, isn't simply how you make VR safe for all people, but how you do it *now*, in the technology's relative infancy. The internet at large, and social media platforms like Twitter and Instagram especially, prioritized growth over safeguards. The result is that these platforms are stuck playing catch-up. Twitter is habitually mired in controversies around abuse and harassment, and last year Instagram unveiled a plan to use artificial intelligence to cut down on harassment in its comments—nearly seven years after it launched.

VR has a rare opportunity to keep that particular horse inside the barn. It's an opportunity too many have squandered. The authors of a 2014 study about sexual harassment in online gaming wrote that "sexism and its expression may be driving women away from many networked games or forcing them into silent participation rather than active engagement." Not only would such a fate be a crushing blow for inclusion and parity in VR, but it would threaten the technology as a whole.

If developers and companies are able to anticipate and even preempt this kind of toxic behavior, that fate can be averted. If they give users the right tools to empower themselves, if they're proactive about building a positive community, and if they're willing to enact—and *enforce*—policies that keep social VR welcoming to all, then presence will have a chance to become everything it promises.

Harassment and toxic behavior are the most pressing issues in social VR's early days, but they're far from the only ones. The dynamics and effects of the technology continue to attract keen interest from cognitive scientists, psychiatrists, and other academic circles; after all, VR gives them a way to cook up just about any scenario they can imagine.

Case in point: Recently, an interdisciplinary group of researchers at the University of Vienna set out to see how being socially excluded in VR affected people's mental states in the real world. Each volunteer entered a virtual environment, a sunny park, where they were invited to play a ball game with two other people. (Beforehand, some of the volunteers had been placed in a waiting room with another person and told that they'd be playing a VR game with that other person; the other half were told they'd be playing with computer-generated people. But in reality, *all* the people volunteers met in VR were computer-generated.) For half of the volunteers, the ball got thrown to them about a third of the time, just as expected. For the other half, though, the game turned into a middle-school freeze-out: after a minute or so, the virtual players simply stopped throwing the ball to the volunteer and wouldn't respond if the volunteer asked why.

The immediate results weren't much of a surprise. The volunteers who were excluded later reported significantly more anger, sadness, and uncertainty than the ones who were included—and those who thought they'd been playing with a real human were even sadder. Conversely, of those volunteers who had been included for the full game, those who thought they were playing with real people reported significantly higher self-confidence than those who knew they were playing with computer-generated partners. Already, it's clear that social VR—experiencing presence with real people—can induce a stronger emotional response than simply interacting with a game character or another computer-generated construct.

It gets more interesting, though. After each volunteer had gone through the VR experience, an experimenter dropped a pencil in front of them; the researchers kept track of how long it took volunteers to react and then to pick up the pencil. Those who had been excluded took significantly longer to do both. That antisocial reaction had been borne out by past research; what was new this time around was that those who thought they'd been ostracized by humans took significantly longer than the others who had been excluded. Whereas with "computer people," they might have been willing to chalk it up to a technical glitch, the sting of being rejected by virtual humans was as visceral as if it had happened in real life. (Conversely, an earlier study found that people who had become "superheroes" in VR, flying through the air and saving a child, were more likely to help pick up pens in a similar situation—just as alienation can be induced, so can altruism.)

The way we treat each other matters. This holds true online, offline, and maybe most of all in VR, where anonymity and presence create a sort of embodied omnipotence. How burgeoning social VR platforms reckon with that power remains to be seen, but even small failings right now can help shape developers' thinking and design decisions in ways that have far-reaching effects. As new realities become an ever-increasing part of our digital lives—which themselves are merging with our "real" lives to the point where the distinction will soon become moot, if still technically accurate—the architects of those realities take on the burden of avoiding, and even redeeming, the mistakes of those who came before them.

Will they do those things? I can't say. I've heard a lot of people say a lot of insightful things, things that evince an earnest desire to create a space in which people don't feel excluded or ostracized, let alone threatened. But companies that want to grow fast sometimes forget to grow smart—and sometimes the smartest companies don't grow fast enough to stick around. So let's take a snapshot.

## AND NOW, A BRIEF SURVEY

No, not that kind of survey. Since we've been talking about the astounding power of social presence, it seems like a good time to take a step back so you can take in the social VR landscape as a whole. At the end of 2017, there are fewer than ten persistent social VR worlds, each with its own user communities, its own features, and its own challenges. There are dozens more VR experiences and apps that feature social presence—multiplayer games like *QuiVr*, or Hulu's VR app, which allows you to connect with friends and watch the streaming service's offerings on a 2-D screen embedded in a virtual world—but these are the ones that focus on creation and interaction:

**Altspace:** I'm gonna go ahead and guess you get this one by now.

**Anyland:** This is "open-world" VR in the most literal sense. The creation of a two-person team, Anyland puts all the tools of creation and communication in users' hands. When you begin, you're just a pair of hands, but you use the game's sculpture tools to fashion yourself a head (your avatar) and your home—then you find other people to spend time with. It may not be as welcoming as many other platforms, just by virtue of its degree of difficulty, but it has a small, ardent community.

**Bigscreen:** This social VR app is built on the power of . . . well, a big screen. Instead of sitting around a campfire, invite other users to play games or watch movies sitting around a giant version of your computer monitor. "We aren't building the 'metaverse' and our goal is not to build a social network," the company's founder wrote in a post on the company's blog. "Instead, we aim to build a platform that enables people to use existing content, apps, and games in VR, and to socialize and hangout in a shared virtual space with their friends and co-workers."

**High Fidelity:** If the name Philip Rosedale sounds familiar to you, you've probably spent some time in Second Life. Rosedale's company, Linden Lab, founded the much-ballyhooed online world in 2003; Rosedale left the company in 2009 and founded High Fidelity a few years later. It's one of a few social VR platforms that aims to become as big as the internet itself, inviting users to create VR worlds as sprawling and detailed as those they made in Second Life. High Fidelity is still in "beta," meaning it hasn't been officially released, but it's notable for allowing user avatars with real-time facial expressions and gaze-tracking—functions that aren't yet available with a headset, but ones that users can enable with extra sensors. In other words, High Fidelity is an alpha nerd's paradise, but maybe a little daunting for everyday users.

**Rec Room:** One of the most popular social VR platforms at the moment, Rec Room is built around the idea of group games and activities. It has a cartoonish aesthetic but also an immediate and palpable sense of humor: people high-fiving creates a cloud of confetti, and if you want to join a party with someone, you simply fist-bump. We'll spend much more time in Rec Room in a couple of chapters.

**Sansar:** The company that Philip Rosedale left, Linden Labs, still runs Second Life—but it also now has Sansar, a VR equivalent that's built from the ground up. Much like High Fidelity, it allows users to design and create stunning worlds via a sophisticated game engine and envisions being as expansive as the internet itself. Also much like High Fidelity, though, it's still in its early stages.

**VRChat:** Despite a somewhat steep learning curve, VRChat has a robust user community, many of whom design avatars that feel like a copyright lawsuit waiting to happen: a walking Intel microchip, well-known video game characters, Iron Man, even Jared Leto as the Joker from *Suicide Squad*. It's aggressively silly, and has become hugely popular in recent months, but is less structured than a platform like Altspace or Rec Room.

**vTime:** Another early entrant in the social VR world, "sociable network" vTime focuses strictly on conversation: you design an avatar and then meet up with other users in environments that range from bucolic (a river in a mountain valley) to fantastic (perched atop the international space station) to pedestrian (a conference table in a high-rise office—yes, you can have a work meeting in there). This is the rare social VR app that works on just about every VR headset there is, even cheapies like a Google Cardboard.

**TheWaveVR:** Part social VR, part music festival on demand, TheWaveVR swaps out activities for virtual concerts—putting on shows in environments called "waves," where users can congregate to listen, dance, or just gawk at the often psychedelic visuals.

There's one social VR platform that I left off this list, though. It's a biggie—in fact, it's probably the first thing you think of when someone says the word "social." It's Facebook. And as it turns out, Facebook is thinking about VR and social media in a much different way than any of these other companies are. Its approach, and what it means for VR and relationships, is best reserved for another chapter. So let's head there now.

# 6

# THE STARRY NIGHT
# THAT WASN'T THERE
## SOCIAL MEDIA, INTIMACY, AND
## THE MEMORY OF EXPERIENCE

W HEN THE WORLD comes into focus, I'm standing at a table, a Facebook employee named Max to my left. Kind of. Max is actually standing in a nearby room with his own headset on; who I'm talking to is Max's cartoonish avatar. He's got brown hair, a beard to match, and a friendly smile. "Well, hey!" he says. "Let's get you started. See that dock floating in front of you? Reach out and touch where it says 'Photos.'"

First, I look down at my hands. Thanks to the Oculus Touch controllers I'm holding in real life, I have hands in here as well. Granted, they're translucent blue, and their fingers don't register every minuscule wiggle of my own, but when I turn my left wrist toward myself I can see the time on my virtual watch. (On this particular sunny day it's 3:50 in the afternoon.) The buttons on the controllers know whether or not a finger is on them and then translate various combinations into VR hand gestures; I can shake my loosely opened hand to wave at someone, make a fist or a thumbs-up sign, or pick something up.

Or, by extending my index finger, I can point. So that's exactly what I do, reaching out toward a blue rectangle that says "Photos." As soon as my finger gets to the box, it glows, and my controller buzzes ever so slightly—a tiny bit of sensory feedback that increases the illusion of having made physical contact. Up pop six Facebook photos of me. This is the first step in a long journey for the biggest social-media company in the world, and it's been a long time in the making.

## THE FUTURE ON YOUR FACE

Mark Zuckerberg decided to buy Oculus in 2014 because he saw virtual reality not as an entertaining escape, but as a human connector. "This is really a new communication platform," he wrote in—what else?—a Facebook post announcing the purchase. "Imagine sharing not just moments with your friends online, but entire experiences and adventures." And that's exactly what Facebook set about planning for. Two years later, Zuckerberg took the stage at the Mobile World Congress in Barcelona and announced that Facebook had established a team dedicated to building "VR social apps." What that meant was a mystery to most of the people there; Zuckerberg offered no elaboration. It would be nearly two months before anyone saw what Facebook had in mind.

F8 is Facebook's annual "developers' conference," tech-industry code for a company-hosted multiday event that's heavy on back-patting and future promises. The second day's opening keynote featured Facebook executives detailing the progress the company was making on various fronts. Mike Schroepfer, Facebook's chief technology officer, spent the better part of a half hour alone onstage at San Francisco's Fort Mason talking about artificial intelligence and virtual reality, but then things took a hands-on turn. "It's always hard to get a sense of all this stuff when you're talking about it and you're showing videos," Schroepfer said, holding an Oculus Rift. "So I thought what we'd do is go into VR together and see what this is like in the real world." He slipped the Rift over his head.

A giant screen behind him showed his perspective inside the headset: standing on a square black mat, a nondescript gray floor fading away to the horizon. This sort of bare-bones minimalism is a hallmark of VR demonstrations—the better for you to concentrate on the stuff directly in front of you. What was in front of Schroepfer, though, was another person . . . or at least another person's head and hands. The head wore glasses and had a goatee and receding hairline: this was Mike Booth, Schroepfer's colleague who was heading up that "VR social app" team that Zuckerberg had mentioned. "Hey, Schroepf, how ya doin'?" Booth asked merrily, waving a hand in greeting. "Sorry I couldn't make it to F8—I'm stuck here at Facebook HQ."

Booth started showing Schroepfer some of the objects that littered the black mat; this was a "test bed," he said, a way for Facebook to investigate how people could interact in VR. The first thing he held up looked like a crystal ball with a tiny scene inside it; it was distorted like a fisheye lens, so you couldn't quite make out what it was. "Why don't you grab that and slide it right onto your face?" Booth said. Inside the convention center, Schroepfer reached out his right hand, which was holding a controller, and took the small sphere from Booth. When he brought it closer to his face, it seemed to bloom outward and surround them. Now everyone could see what the scene in the ball had been: a train station, with a long arched glass roof. It was a 360-degree photo—that the two men were *standing within.* "Welcome to St. Pancras Station, London!" Mike Booth said.

The demonstration turned into a virtual travelogue, with the two men picking up ball after ball and bringing the duo into each 360-degree photo held within: Piccadilly Circus; a giant hangar where one of Facebook's enormous internet-beaming drones (a topic for a totally different book) was being built; and finally, Westminster Bridge, straddling the Thames River.

Schroepfer reached out and pointed at two tourists. "Hey," he said, "they're getting a selfie in front of Big Ben. *I* never got a selfie in front of Big Ben." (Scripted patter at developer conferences is as painful as it is at awards ceremonies and in airline safety instructions.)

"Ah, I may have a solution for that," Booth said, bending down to pick up another item from the mat: a slender stick with a screen attached to one end. "Another experiment we've been trying is a virtual selfie stick!" The audience laughed—and then, as Schroepfer took the selfie stick and angled it toward himself and Booth, they began to applaud. The magic show wasn't finished, though. A few colored pens were also sitting on the mat; the two men used them to draw neckties and then stick them onto each other's avatar. Properly accessorized, they took some virtual photos—Schroepfer mugging and giving a thumbs-up, courtesy of his hand controllers—and wrapped up the demonstration. It was barely more than four minutes long, but it put other social platforms on notice: Facebook wouldn't just be connecting you with people in prerendered environments; it would be allowing you to travel together.

In fact, that was just the beginning. A few months later, it was Mark Zuckerberg's turn to wow a developer conference. This time, the conference in question was Oculus Connect, an annual event held by the VR company to invigorate the community it depended on more than any other. (After all, people buy headsets only if there's fun stuff to do with them.) Zuckerberg's demo was in many ways a more polished, advanced version of what Booth and Schroepfer had shown off. The avatars, while still cartoonish, were much more lifelike; Zuckerberg's made him look like NSYNC-era Justin Timberlake. Instead of 360-degree photos, he and his colleagues horsed around in video environments, watching sharks swim around them and checking in with a rover on the surface of Mars. This version of Facebook's VR world featured new toys to play with, including playing cards and a screen that let the users watch 2-D videos within VR. When Zuckerberg took the now-obligatory VR selfie, he posted it to his real Facebook feed.

Yet, the demo had a distinctly pedestrian undercurrent. Zuckerberg and his co-workers visited the bottom of the ocean and Mars, but also teleported into his own office, and even his living room at home—where his dog, Beast, stared obliviously into the middle distance. But then things got weird. A ringing-phone noise alerted Zuckerberg that he was getting a call via Facebook's messaging tool; when he looked at his watch in VR, he saw that the call was from his wife, Priscilla. He reached out with a finger and selected "answer," which launched a smartphone-shaped screen in midair. In it, he saw Priscilla in real-life form, while she, looking at her phone, saw him as a VR avatar. ("Why do you look like Justin Timberlake?" she oh-so-spontaneously asked.) All of it was calibrated to make VR feel not simply like escapism, but part of your day-to-day life.

And more important, that demonstration created a wormhole between the real world and VR. Virtual reality has always been about erasing the frame; Facebook's project, by allowing smartphones to tunnel into VR without a headset, promised that VR wouldn't be a gated neighborhood separated from the rest of the internet.

## INNER SPACES

It took another few months, but in 2017 Facebook finally turned those demonstrations into something people could use—and Spaces, as it's now known, is something wholly different from the social VR platforms we've already visited. To use a universal metaphor: VR is the singles bar, and the shared activity is the alcohol. You enter VR to meet new people and create connection, and you create those connections in the course of playing Dungeons & Dragons, or paintball, or any of the other things that the platform has built.

If Altspace is a singles bar, though, then Spaces is (to use a similarly dated–sounding metaphor) more like a coffee klatch. You're there to deepen a connection you already have. When you launch it from within your headset, it logs you into Facebook—the same Facebook everyone you know uses regularly; the same Facebook you've already populated with all the information that Spaces now can serve back to you. Like, for instance, recent photos of yourself that you can use to create an avatar.

And much like meeting close friends for coffee, what you actually *do* in Spaces is almost immaterial. Facebook places a premium on the participants rather than the activity. You're not there to play disc golf, you're there to hang out. You can choose your surroundings—any 360-degree photo or video you can find on Facebook—but beyond that, entertainment is a bit scarce. There's a marker you can use to draw 3-D objects, which you and your friends can interact with: Draw a hat and put it on their head! Trade cartoonishly villainous mustaches! A mirror is there so you can gaze upon your majestic cartoon countenance. (Don't act like you wouldn't do it.) And of course there's that selfie stick, which you can use to take a VR photo of yourself and your friends and then post that photo to your real-life Facebook feed.

Make no mistake, this is all silly stuff. But that silliness masks a deeper truth: VR in Facebook's view isn't meant to create relationships, but to deepen existing relationships. Relationships are built through shared experience. That's what moments are, that's what the tapestry of intimacy is woven from. Spaces gives you a palette of new experiences, which in turn can lead to more moments and intimacy than you might otherwise have shared with the other person. "It's not even peanut butter and chocolate," says Rachel Rubin Franklin, who heads Facebook's social VR efforts. "It's some magic combination of taking presence in VR, matching up with people that you already care about, and then saying, okay, what would you want to experience together?" (Before joining Facebook in late 2016, Franklin spent years at the video game company Electronic Arts, including a stint overseeing *The Sims 4*.)

What Spaces really comes down to is treating identity as a

known quantity. Throughout its forty-year history, the internet in all its forms—from email to dial-up bulletin boards to the Web to Twitter to Snapchat to Flipsock to Gluzzzz (okay, I made up those last two, but it's only a matter of time, right?)—has hinged on your ability to present yourself however you desire. You might be QueenJambalaya on a Creole cooking forum and sexytime42@hotmail.com on the throwaway email account you use to sign up for newsletters and coupons, but those identities aren't mutually exclusive. You're simply accentuating a dimension of yourself that's suited to the purpose at hand.

That's not to say that anonymity can't accelerate intimacy, of course. You don't need decades of sociological research to tell you that using an alias can lower your inhibitions; anyone who's ever been a regular in a chat room or on a subreddit knows how real-life strangers can unburden themselves to one another online. However, that intimacy often arises only as a *condition* of anonymity: just because you know about PhilTheThrill's childhood traumas or commitment issues doesn't mean you'd be as vulnerable with each other if you met in real life. In fact, meeting for coffee would likely be more awkward than it would if you were simply real-world acquaintances—you'd both be preoccupied trying to reconcile the person you're talking to with the "person" you know online, a process often complicated by the fact that, for example, PhilTheThrill's real name is Greg.

While online and IRL relationships are more alike than they are different, they evolve along similar but separate tracks. A real-world relationship tends to accrue trust and intimacy slowly, people's innermost selves coming out over time. Meanwhile, because anonymity breeds candor, digital relationships can skip those early steps, or even invert the process: you can know a person's deepest fears and fantasies before you learn their mannerisms, or even how they sound when they laugh.

What you know about someone, and when you know it, also determines what one person projects onto the other. When you get to know someone online, even if you've seen a photo of them, your mental image of them is still speculative: you can't know what sort of real-world chemistry the two of you might share (or lack). And similarly, when you meet and get to know someone IRL, you're doing the same thing in reverse, just imagining their personality and psyche rather than their appearance. Both ways, as anyone who's taken an online relationship offline may know, can lead to disillusionment. (This is the idea behind online dating sites like Match and eHarmony: online honesty + initial online conversations + profile pictures = a more genuine connection than you might find chatting someone up at a bar or party, without the blind uncertainty of nameless online communication.)

The early days of social VR, though, created a third track of relationship, one that exists somewhere between the other two. Most VR platforms allow for the same anonymity that has defined so much of the internet: choosable usernames, say, or customizable avatars that can look either like you or unlike you. However, presence means you're not just being honest from behind a keyboard, but as an embodied identity, sharing a space with other embodied identities. You feel like you're there with someone else; for all intents and purposes, you *are* there with someone else. That can often lead to the same halting progression from shyness to acquaintance to friendship that marks real-world relationships. Just because you're in VR, in other words, doesn't mean you're ready to bare your soul immediately.

Which brings us back to Facebook Spaces. By marrying the

promise of virtual reality—the magic of sharing a vividly experiential activity—with the prebaked intimacy of an existing relationship, Facebook hopes to make social VR an attractive prospect for all the people who might not thrill to the idea of approaching strangers. There's no need to navigate the early stages of interpersonal comfort, as you might in Altspace or another platform. This is Facebook; you're with someone you already know. You could be chatting about something as pedestrian as the day you had at work, but doing it in—and I'm just looking at the most recent 360-degree video I have saved on Facebook, to give you an example—Knob Lake, Alaska. There you are, talking with your sister who lives three states away, or a college friend you haven't seen in a year, while the lights of the aurora borealis flicker in the northern sky and the dusty gleam of the Milky Way stretches away into the distance.

But here's the thing. The next time you think back to that conversation? The next time someone asks whether you've seen your college friend recently, or how your sister is? You'll know exactly when and where you were when you spoke. But the answer won't be "in VR." It'll be "in Alaska, on the starriest night either of us has ever seen."

For that, you can thank your brain.

## MISTY WATERCOLOR MEMORIES (OF THE WAY VR)

We know presence can affect our emotional responses; we know it can affect our physical responses. It stands to reason, then, that VR can affect our memory as well—or at least one particular type of memory.

Memories can be divided into two major types: implicit and explicit. Explicit memories are those that you can express in words—and those can also be divided into two major subtypes. (I promise, this is as specific as we're going to get.) Explicit memories are either things you *know*, called "semantic memory," or things you have *experienced*, called "episodic memory." Episodic memory is sometimes referred to as "autobiographical memory," since it's concerned with the events of your own life. (Your knowledge of another person's life falls into the realm of semantic memory, since it's a collection of facts rather than actual experiences.)

Since VR is an experiential technology, we're going to worry about only episodic memory right now. When you put on a headset, you're effectively experiencing two different things. At the purely corporeal level, you're standing or sitting with a headset on, and possibly moving your head and hands around. But the virtual things you're seeing, the virtual things you're doing, become the true episodic memories.

We see this a bit with so-called first-person video games, in which you interact with the game's world from the perspective of your own character. If you've been playing Scrabble for an hour, your memories will be of the tiles you drew, the words you made, where certain words sat on the board, your opponent's reaction when you laid an *X* down for a two-way triple-letter score. However, if you spent an hour playing a first-person shooter, your memories will be very different. You'll remember crouching in a warehouse waiting for an opponent to run in; you'll remember dodging across a courtyard while shots whistled by; you'll remember jumping off a roof to surprise someone below. Obviously, those aren't things you *did*; what you actually did was press a button to crouch or push a thumbstick in a particular direction to run. But because the game presents itself in relation to your viewpoint (what's called "ego-centric spatial coding"), the crouching and suspense and dodging and adrenaline are your experiential memory.

That's part of what's happening in VR—you're experiencing this virtual activity from the perspective of a person experiencing the real version of that activity. The sights and sounds, and in some cases the tactile sensations, match up with what they would be if you were actually doing the thing you're doing virtually.

In fact, VR's ability to do that overturned some fundamental thinking about mnemonics, or the study of memories. This dates back to the mid-2000s, when some memory researchers at Duke University had codified the idea of "laboratory memory": brain scans of people who tried to remember things they'd been shown in photos, versus those in photos they'd actually taken, showed lower activity of the hippocampal region, which is involved in autobiographical memory. But it wasn't just the hippocampus that was involved; laboratory memory also showed lower activity in the regions controlling self-referential processing and visual/spatial memory.

Last year, a trio of German psychologists set out to challenge that idea. In their thinking, if laboratory memories were treated as less strong or pure than organic autobiographical memories, then the adoption of VR in experimental psychology might be threatened. Clinical researchers were already using VR for therapeutic purposes, but the Germans saw huge promise in VR's ability to deliver perfectly tuned simulations of real-life experiences; proving that VR could induce reactions and memories indistinguishable from real-life memories would help validate the use of VR in more fields, including mnemonics. "It has to be clarified," they wrote, "whether VR is merely a fleeting . . . illusion or if the brain accepts the VR experience as an imposed alternative reality on a larger time scale."

The first step of the experiment, oddly enough, was linguistic: the researchers started by getting rid of the terms "episodic memory" and "autobiographical memory." There simply wasn't a consistent enough definition for either, they said. Some studies used the terms synonymously, whereas others treated episodic memory as merely one ingredient of autobiographical memory, along with self-reflection and other processes. Instead, they decided to use the terms "participation based memory" (PBM) and "observation based memory" (OBM). A PBM was richer than an OBM, and more personally meaningful to someone—and given that complexity, the researchers said, the brain would take longer to retrieve a PBM than an OBM.

As for the experiment itself, it revolved around a long 360-degree video of a motorcycle ride in northwestern Germany. Half of the subjects watched a regular 2-D video of the ride on a fifty-five-inch computer monitor with their hands on the table in front of them. The other half sat on a piano stool and watched the video in a VR headset, their hands on a pair of bicycle handlebars. Two days later, all the subjects were shown a parade of images, each for two seconds at a time; some were taken from the video of the motorcycle ride, others from a different ride in the same area. For each image, subjects pressed a button to signify whether or not they recognized the scene.

By now, it should come as no surprise to learn that the VR group thought their experience of the "ride" was significantly more realistic than those in the video group. The VR group also performed more than twice as well in the memory test—but perhaps most interesting, it took them almost a half second longer on average to respond to each image. In other words, the memories of the VR "ride" behaved like those of a person who had actually participated in the motorcycle ride. These were PBMs, not OBMs. "VR represents a new category of immersion," the researchers wrote. "Virtual reality actually deserves the term reality—with all its prospects and risks."

## DOING THE DO

Intensifying VR's ability to create strong memories is the fact that sometimes you're *doing* the thing you're "doing." Rather than simply being a voyeur, you're physically participating in the virtual activity. Embodied presence—bringing not just your head and your hands, but your body into VR—strengthens memories in a number of ways. Actually walking within a large-scale VR environment, rather than using a joystick, helps people navigate more efficiently, and helps them create a better mental map of a space. In VR, you remember more of what you see when controlling a car's speed than simply being a passenger. (But not when you're actually using a steering wheel—interestingly enough, that demands enough focus that your memory of the surroundings suffers.) VR that encourages embodied presence holds promise for improving episodic memory in seniors, many of whom become less mobile as they age.

Facebook didn't create Spaces to improve people's memories, obviously. But the tools available in Spaces—the selfie stick or the marker you can use to draw 3-D objects—are about harnessing that creative freedom to create *new* memories. "We could give you one activity to do, and that would be fun for a while," Rachel Rubin

Franklin says, "but if we give you something that's open-ended, then you're going to figure out what to do with it."

When Franklin joined Facebook, the social VR group had already been exploring various activities: tower-defense board games, music-creation games, even the ability to design a doll-house and then teleport into it. But the more people had to do, the team found, the more they concentrated on the activity—and the less they concentrated on each other. The group formulated a litmus test for what Spaces would include: Does it facilitate social interaction? As in, is it going to make my relationship with you stronger, better, more memorable? "If that's not happening," Franklin says, "then it doesn't belong in here. At least not now."

But there's also something else about the tools that the social VR team built into Spaces. Thanks to hand controllers, you're physically using them. When you hold the "marker," you're really drawing the shape of a hat in the air. When you extend the selfie stick and position it just so, so it captures you *and* your friend *and* the panda behind you, you're really doing that in the real world. You're not engaging with them passively, and you're not con-trolling them using an alienating metaphor like a keyboard or a joystick—you're holding something and you're moving it in space. The sights and sounds you experience in VR might help create memories; the activities really cement them. The more you do in VR, the stronger your memory becomes.

Besides, that selfie stick is just a recording tool. Facebook's long-term plans with VR aren't necessarily about sharing that selfie, they're about establishing Facebook as a place to go for the experiences.

At the beginning of 2017, Facebook published some of its inter-nal research about the potential of social VR. Rather than set up its

## A Word About Research

People love studies. Just the word sounds official, right? As you probably know, though, the word is about as official as the word "article." Just about anyone can write one, and publish it just about anywhere. And just like places that publish articles, the places that publish studies have varying reputations, with varying degrees of rigor that they demand of said studies. Read a story in the *New York Times* or, oh, say, *WIRED*, and you can be assured that it's been copyedited and fact-checked to be verifiable and truthful. The same goes for the faith that studies appearing in certain journals are methodologically unimpeachable: double-blind, thoroughly vetted, replicable, and everything else. But not every media company is the *New York Times* or, oh, say, *WIRED*—and not every study is unimpeachable.

I say all this for a reason, though, and that reason has to do with the huge body of research that doesn't fit into either category, because it's privately funded. As a journalist who has to think about this kind of thing, I don't mind saying that dealing with that kind of research can be a little fraught: if a study about a given subject is funded by an organization or company whose commercial purpose is tied to that subject, then you might be looking down the barrel of a conflict of interest. But a lot of companies fund their own research—and often have way, way, *way* more money to devote to research than a university lab has. So just because Gatorade's sports science lab does a giant study about hydration in endurance athletes doesn't automatically mean the findings are bunk. And just because Facebook/Oculus—or Microsoft, or Google, or any large company—has funded its own studies about VR doesn't necessarily disqualify said research. (What role that research plays in the larger scientific community is another question, and one that's probably big enough for its own book.) ∎

own study, the company had used a third party: an "applied neu-roscience company" called Neurons Inc. that attempts to quantify people's reactions to various products and businesses. For example, smartphone company Ericsson once commissioned a report from Neurons Inc. about the effects of streaming delays on consumers. The agency measured the eye movements, brain activity, and pulse of volunteers who were watching streaming video on smartphones and ultimately discovered that buffering and lag were significantly more stressful than waiting in line at a store, and even slightly more stressful than watching a horror movie.

Facebook didn't want to measure the effects of technical hiccups, though. It wanted to know how people actually acted in virtual reality, and how they felt about it. So Neurons Inc. found sixty people who had never met and divided them into thirty pairs—half of whom were told to sit in person and have a short conversation and half of whom were given the same assignment, but while wearing VR headsets. (The virtual environment looked like a train car with art deco touches, and the two participants sat on facing leather seats. Each person was represented by a reasonably accurate avatar, essentially a good-quality video game character that matched the participant's hair, skin tone, and general build.)

The researchers asked everyone to begin with small talk and then move on to "discuss more personal topics." All the participants wore sensors that measured their brain activity; most of the people in the VR group had never used the technology before.

Not entirely surprisingly, most of the subjects in the VR group responded positively to the experience: participants guessed that the conversations had lasted thirteen minutes, when in reality they lasted twenty. However, what *was* surprising was how strongly introverts reacted to VR. After the VR experience, more than 80 percent of introverts—as identified by a short survey participants took beforehand—wanted to become friends with the person they'd chatted with, as opposed to less than 60 percent of extroverts.

This carried over into more quantifiable realms as well. One of the properties Neurons Inc. measured was "engagement," which was a combination of EEG results that showed high arousal and positive motivation. (High arousal and *negative* motivation could indicate fear, while low arousal could point to boredom or dislike.) Introverts, on average, displayed higher engagement during their VR experiences than when meeting someone in person. Further, for people in the VR group, their "cognitive effort levels" lowered as time went on, suggesting that it became easier for them to chat, even as the conversation veered into personal territory.

It's not that VR, for these people, had some kind of magical stress-reduction properties. (To the contrary, in fact. VR's ability to induce anxiety by placing people in socially stressful situations, such as a blind date or a job interview, or exposing them to their phobias, has made it a fertile area of exploration for exposure therapy, even proving to be more effective than real-life exposure therapy.) It's more that it engendered a new kind of intimacy— one that's able to flourish without stakes. In a short video that Neurons Inc. made chronicling the experiment, a pair of women talk about the experience they had meeting as strangers on that VR train car. "It allows you to be more open because the other person isn't, like, staring at you," said one of them. "You're not trying to figure out if the other person is judging you."

Is the other person *actually* judging you? Probably not. But self-consciousness isn't really a stickler for accuracy. And if VR eases that self-consciousness by making the interaction seem somehow less real, all the better for people who might experience social anxiety. Just ask another one of the volunteers in that experiment, a man whom the video captures opening up to his new VR acquaintance about the past: "I regret not telling him that I loved him," the guy says, headset covering his face. (And as you might be thinking right now: yes, that means that VR might actually lead to *over*sharing. Choose your VR buddies wisely.)

But the volunteers aren't the only people who appear in that video. So do some Facebook employees—including Rachel Rubin Franklin, who delivers what might be the most telling comment yet about Facebook's long-range plans for VR. "To see that people are responding so well means we can bring billions of people closer together," she says. "It's an exciting future for us." In other words: while you might be able to connect only with your preexisting friends right now, that's likely to change.

After all, Facebook may have started out as an actual book of faces, but a decade later it's far, far more than that. It's a media company and a game publisher, a community organizer and a videoconferencing app. VR is going to have tendrils in every one of those uses—and Facebook is going to make sure of it.

# 7

# REC ROOM CONFIDENTIAL
## THE ANATOMY AND EVOLUTION
## OF VR FRIENDSHIPS

I T WASN'T THE DRIVE that made Ben nervous. Driving from his home in Cincinnati to Birmingham, Alabama, takes just over seven hours, and he'd driven longer than that before. Besides, the twenty-four-year-old hadn't taken a vacation in a while, so he was looking forward to the time on the road. His trepidation stemmed more from what was waiting for him at the other end of the trip: his new friend.

In 2016, Ben had decided that he wanted a VR headset for Christmas—a nice one. He already had a desktop computer he used for gaming, so he asked his entire family to try something new that holiday. Instead of each person getting him a present, he wanted everyone to chip in a little bit to help defray the cost of the HTC Vive he wanted. Since Ben was kicking in $250 of his own money, it didn't leave *that* much for everyone to contribute. The Vive plan worked, and by Christmas Day Ben was ready to enter the Metaverse.

What he would do in there, though, was a question he hadn't yet answered. So, like so many computer-savvy twenty-somethings, he consulted Reddit. A subreddit was already devoted to the Vive, and there Ben saw a post about a free game called Rec Room (see page 126). What he didn't know then, of course, was that he was about to encounter one of VR's most notable new worlds: a place that helped people overcome social anxiety, facilitated close-knit friendships, and would eventually inspire him to take a seven-hour road trip.

## WELCOME TO THE CLUB

The Rec Room developers, a tiny Seattle company named Against Gravity, call it a "virtual reality social club," which seems as good a way as any to describe a social VR app built around a gym-class metaphor. When you launch the game, you're in your personal dorm room, with a loft-style bed against one wall and a dresser and mirror against another. Pieces of paper pinned to a bulletin board show you how to use your hand controllers to grab and use items or to teleport around. (Although you can physically walk around in Rec Room, you're bounded by the headset's cable and tracking system. For covering bigger distances, most VR experiences default to a locomotion system in which people use their controllers to specify a point they then "teleport" to. Earlier solutions, which relied on video game conventions like holding a thumbstick to "walk" in a given direction, were much more likely to result in motion sickness.)

You can also select your clothes and customize your avatar, looking in the mirror to admire your getup. Rec Room gives users daily quests, some of which unlock new gear like a snazzy sea captain's cap or a mod-style shift dress straight out of *Mad Men*. The options are remarkably broad, as well as inclusive—you can wear Pippi Longstocking–style braids and a goatee if you want, and having a beard doesn't mean you can't wear a miniskirt. However, there's nothing photorealistic about it.

Leaving your dorm takes you to the locker room, essentially a gathering place for all the other avatars of players who are online in Rec Room. A woman's British-accented disembodied voice fills your ears, cheerily telling you about various things you can do: "when you're ready to go play, follow the . . . arrows to the back of the locker room and teleport through an activity door." Those activities range from P.E. staples like dodge ball and soccer to intricate themed games in which you and other players fight your way through waves of killer robots or medieval baddies. But even if you don't choose an activity, the locker room is littered with points of casual connection: a basketball hoop, ping-pong tables, lounges with chairs.

When Ben first walked into the locker room, he was dressed like the first person who had come to mind when he was designing his avatar: Donald Trump. Blondish nest of hair, suit and tie, the whole thing. "I trolled people a little bit," he says. "I didn't do it too bad, and I only did that for a few days." He didn't make any friends while he was doing that—but he didn't get kicked out either, which is proof that he probably wasn't being that annoying.

## Avatar Design

You've probably noticed by now that there's been a lot of talk about "cartoonish" avatars, but not a lot about realistic ones. First off, making a computer-generated face that moves realistically is tough. Like, *really* tough. Even video games with hundred-million-dollar budgets struggle with it. The way your lips and tongue constantly rearrange themselves according to the words you're saying, the way your eyes crinkle at the corners when you smile, the way light bounces off your skin, even the way your hair moves—each of these physiological elements poses brutally difficult technological challenges for software developers. But even if all of these things were unerringly perfect, there's still the Uncanny Valley.

*The what?*

Great name, right? In 1970 a Japanese roboticist named Masahiro Mori wrote an essay stipulating that the more lifelike a robot is—having eyes that blink, for example, or a metal skeleton covered with something that looks like skin—the greater affinity humans feel toward it. However, there comes a point at which a robot's humanness becomes deeply unsettling. As Mori wrote:

> Recently, owing to great advances in fabrication technology, we cannot distinguish at a glance a prosthetic hand from a real one. Some models simulate wrinkles, veins, fingernails, and even fingerprints. Though similar to a real hand, the prosthetic hand's color is pinker, as if it had just come out of the bath.
>
> … However, when we realize the hand, which at first sight looked real, is in fact artificial, we experience an eerie sensation. For example, we could be startled during a handshake by its limp boneless grip together with its texture and coldness. When this happens, we lose our sense of affinity, and the hand becomes uncanny.

And when something humanlike is moving, rather than being still, the Uncanny Valley gets even deeper. A fake hand making a grasping motion is eerier than a fake hand at rest—and a lurching zombie is far more off-putting than a corpse.

For VR avatars, the Uncanny Valley is incredibly difficult to cross: an otherwise photorealistic 3-D rendering of a human face could be ruined if the eyes aren't absolutely perfect. Most software companies creating social VR, then, opt to steer clear of the valley altogether.

That's not to say that today's avatars aren't accurate or expressive, even in these early days of VR. When Facebook's social VR team was designing how people would look in Spaces, it used a two-part golden rule: you should be comfortable with the way you look, and your close friends should recognize you at a glance. To that end, Spaces examines your real-life photos and creates a stylized abstraction of your facial features, hair, skin tone, and accessories, such as glasses. My avatar in Rec Room looks like me only in that it has a shaved head and a beard; on the plus side, though, with its dozens of options, its wardrobe is *way* better than my Facebook avatar's, which can choose only what color T-shirt to wear.

When it comes to talking, things get a bit more complicated. Trying to make lips move accurately can "get really uncanny," Facebook's Mike Booth has said. Instead, Spaces uses a discrete number of mouth shapes and snaps from one to the other as it hears you talk. Similarly, if you turn your head toward another person, Spaces will automatically shift your avatar's gaze and create the illusion of eye contact. (In all likelihood, the next generation of headsets will actually track your pupils, which will make that eye contact much more responsive and lifelike.)

Rec Room, for its part, does away with features entirely: avatars are noseless, and they all have the same eyes and mouth, though those can change to fit the expression that you *sound* like you would have. If you're laughing or talking loudly, your eyes will shut and your mouth will look like a big, open smile. If you're silent, you'll have a small, pleasant grin on your face.

But your voice isn't the only thing that can change your avatar's expression. Some social platforms also use a limited set of expressions that are triggered by your natural hand actions. In Spaces, if you shake a fist (using a controller, of course), your avatar will look angry; if you raise both hands to your face, your avatar's mouth will open and it'll look terrified. The emotions being conveyed aren't nuanced, but they're more effective than you might think—after all, we're already used to communicating in broad visual strokes. Just look at your phone. Current thinking holds that there are only about twenty fundamental facial expressions. As Elli Shapiro, who helped develop social applications for Sony's PlayStation VR, has pointed out, those expressions map astonishingly well to emoji; it's easy to imagine the tiny texting-friendly version of "fearful" or "happily surprised" or "angrily disgusted." (For those keeping score at home, those would be 😕, 😄, and 😖.) Emoji are the perfect way to think about the current state of avatar expressions, in fact: there's a finite number of them, and they translate pretty much universally.

Currently, social VR in general hews to this cartoonish aesthetic: mouths that move in response to hearing your voice, eyes that periodically blink and look around, features that are charmingly, approachably stylized. As one recent study found, people communicating with VR avatars wearing an "enhanced" smile not only felt stronger social presence than those communicating with "normal"-smiling avatars, but used more positive words afterward to describe the experience. (Findings like these make Altspace seem all the more interesting for its unchangingly stoic avatars.)

> As with everything else in VR, though, all of this is changing, and quickly. A number of companies have developed technologies that allow your real-time facial expressions to drive those of your VR avatar. They might not all cross the Uncanny Valley yet, but they're at least starting to build the bridge. ■

See, Rec Room prides itself on its positivity. Those touches I mentioned before—being showered with confetti for high-fiving each other, or fist-bumping to invite someone to your game? According to Nick Fajt, Against Gravity's CEO and cofounder, such touches are the whole point. The games themselves are simply lubricant. "I don't think we really think of ourselves as a games company," says Fajt. "The area we focus on is creating a community for people from all walks of life, and games are a great way to help a community form."

When you're trying to create a community, though, trolls don't help matters. Even mild ones can drive new users away. So all Rec Room users have a suite of anti-harassment tools at their fingertips—or, rather, at their wrists. If you look at your watch in Rec Room, a little menu pops up that lets you see which of your friends are online or take a picture of yourself, among many other things. You can also report any other player for acting up. As soon as you do, that player gets automatically muted in your headphones, and their avatar fades away to be almost invisible. Other players in the room are prompted to vote whether the bad actor should be removed from the room. Regardless of the outcome of the vote, your report gets logged by the Rec Room system; players who rack up enough reports can be banned.

Taken in conjunction with Rec Room's goofy, self-deprecating aesthetic, the message is clear: be nice. "If you look at the broader internet and social platforms," Fajt says, "these are problems that all of those still deal with, and we're in Year 25 of the consumer internet. If you're not working on them now and you're not taking them seriously now, they're only going to grow."

Rec Room's heavy users, whether through Against Gravity's efforts or merely though self-selection, seem to have taken the suggestion to heart. One of them, Jon Ludwig, lives in Japan with his wife; the twenty-nine-year-old is a lapsed gamer, but it's not uncommon for him to spend four or five hours in Rec Room on a weekend. "I can be shy in real life, but in Rec Room, I want to be the absolute best version of myself," he tells me over Skype. "And the best version of myself is a little bit more outgoing, and a little bit more willing to talk to people and make sure people are having a good time—if they're a little new, even complimenting them. Every time. I just don't even think about it."

The feel-good vibe worked on Ben too. Soon enough, he stopped dressing his avatar as Donald Trump, and he started noticing the same people each time he was in Rec Room.

The last time Ben had a real social circle of friends was high school. In college, he was a commuter student, driving thirty or forty minutes each way to get to his classes; for someone who describes himself as a "loner," that made a social life next to impossible. "If I was going to a party," he says, "I'd have to go home after class to get a shower, get dressed, then go back to campus, where I'm going to have to figure out where to stay for the night—it was just a giant pain in the ass doing stuff like that." So he mostly kept to himself.

Keeping to yourself isn't really an option in social VR, though.

I mean, technically speaking, it *is*, and in any lobby of a social app you'll see people milling around by themselves, but Rec Room all but guarantees that you're going to be talking to people. That's exactly why most of the games are team-based, with cooperative elements; at Against Gravity, Fajt and his colleagues call the guiding principle "structured social."

Eventually, Ben met a young woman named Priscilla. Priscilla had started out in Rec Room even more awkwardly than Ben had; she didn't talk much, and she'd often leave, crying, because she thought people didn't like her. But by the time she met Ben, the twenty-seven-year-old was outgoing enough to tell him to add her as a friend.

In her real life, Priscilla is a successful sports artist, creating incredibly detailed pencil drawings of University of Alabama football players. She's also, in her own words, a "hermit." She grew up in a tiny town in Alabama, but for the past four years, she's worked from home and hasn't ventured out much beyond that. "I just go to the post office," she says, half joking, when describing her non-virtual social life. "As far as, like, having actual people that I talk to almost every day? No."

But now that Priscilla and Ben were friends in Rec Room, they could invite each other online and into specific games, and before long the two were hanging out virtually five nights a week. There were others too. All told, a group of anywhere from two to fifteen people began congregating almost daily. At home, they might a have drink or some weed nearby, while inside their headsets, they'd play paintball or one of the cooperative quests, then reconvene in a private lounge for some drunken 3-D charades and soul-baring.

As the year wore on, the group's closeness spilled over the edges of VR. Regulars congregated in any number of online spaces, from a Rec Room–devoted subreddit to a dedicated space on the chat platform Discord; many of them followed each other on Instagram using Rec Room–specific accounts, some of which never show the people behind the avatars. "A lot of the people in VR have a similar social anxiety," Priscilla says. "I think that's why we connect so well—and why it's, like, way more special than anything else."

Ben agrees. "Everyone who has VR—at least people who actually spend a lot of time in a game like this—they're trying to find something in the virtual world," he says. "I'm like that as well. It's great that we have a medium that we can actually go in, have something as intimate as this."

Priscilla and Ben took things one step further and started texting each other. Then they hit on an idea. Priscilla had been wanting someone from Rec Room to visit her in Alabama, and Ben was more than willing to tackle the drive. "I figured, hey, I can do that for a friend like I haven't had in years," he says.

So he did.

## WHEN AVATARS MEET

Taking online relationships offline is nothing new; between 2005 and 2013, one large study found that more than a third of all marriages began online. Thanks to social presence, though, VR represents an entirely novel variation. There's simply no other communications medium—texting, emails, chat rooms, instant messaging, social media, even FaceTime—that can place two people in the same space in such an embodied fashion. In the case of Rec Room, you can get to know someone's mannerisms and unique quirks and become real friends with them—even if all you've seen of them is an avatar with no nose.

By the time Ben knocked on Priscilla's door, he'd seen her nose, and she his: they'd traded photos, which had forced Ben to take his first-ever selfie. "That picture you sent me was just not working," she says to him now, laughing about his awkward pic. It's the next-to-last night of Ben's trip to Birmingham, and we're on a Skype video call.

While I'd spent time with both of them in Rec Room—generally getting my ass kicked in paintball—before this I knew them only as their avatars: Priscilla as a dark-haired woman who was partial to a police officer's cap, and Ben (or Blitzkrieg, his Rec Room username) as someone who wore an old-timey sheriff getup and had a big blond beard . . . along with a red cape and a top hat, because why the hell not? In person, though, they're perfectly run-of-the-mill twenty-somethings. Priscilla's pretty, with big dark expressive eyes to match her hair, and Ben is tall and broad, with wavy blond hair and a mild demeanor.

Talking to them as flesh-and-blood humans feels more "real" than VR by most conventional measures—but, to my surprise, it's also very much not. Much of that is due to that weird alienating gazelessness so endemic to video call options like Skype and FaceTime. Instead of looking at your phone or laptop's camera, you look at the person you're talking to, and they do the same—the net effect being that everyone's eyes are slightly averted. In VR, our avatar's eyes *are* the camera; to look at someone's face when you're talking to them is to give them not just your undivided attention, but the *appearance* of your attention.

There's a sense of distance that pervades our video call, despite the fact that their personal mannerisms persist. Ben still has a tendency to rephrase my questions when he's talking, like a beauty pageant contestant, and Priscilla still tends to pause to gather her thoughts, and "I mean" is still her favorite phrase—but the social conventions of the real world make the conversation feel like an interview, instead of just organic conversation. When we've hung out in VR, the ease and intimacy between them has been palpable; once, in Rec Room, while Ben was talking to me about his job, Priscilla grabbed a marker and some yellow sticky notes, drew a bikini top and cleavage, and affixed them to Ben's chest. It was funny, but I was struck more than anything by how comfortable the two seemed around each other. That's not on display right now; they're sitting in two chairs, and there's little physical interaction between them.

Of course, some of that might be attributable to the will-they-or-won't-they vibe they're giving off. When I first learned that Ben was heading to visit Priscilla, I assumed that his trip had an ulterior, if obvious, motive; the way he so carefully avoided nonplatonic references in our VR conversations only bolstered that idea. (At some point, when unsure of someone else's feelings, we've all opted for self-preservation through plausible deniability.) And our Skype conversation now has steered clear of any mention of romance.

When Ben first got to Priscilla's place, he tells me, they hugged each other and she showed him around; then they watched comedy movies and fell asleep on the couch. (Priscilla wanted to watch *The Notebook*, but it wasn't streaming.) They went to Buffalo Wild Wings, and Ben persuaded Priscilla to go hiking. Neither says anything about . . . anything. But the real world betrays them: they're both blushing more than their avatars ever did. They're clearly not going to tell me anything, so I ask whether they had talked about the possibility of romance.

"I mean, it would be nice," Ben says, laughing. "But I didn't set expectations in my mind for really anything on this trip. I was just taking a vacation, hanging out with a really nice friend."

So . . . have things moved in that direction?

Ben looks at Priscilla: "I'll let you handle that one."

"Can we tell you later?" Priscilla says. As soon as the Skype call ends, she sends me a flurry of texts:

> ok so here's the thing about that question you asked about the romance

> Blitz had feelings for me for some of the time before coming here that he ended up telling me

> I had a crush I was getting over so I didn't actually reciprocate those feelings until his arrival here

> so long story short, we kissed:p

It's sweet, and it's innocent, and it's just like real life. It's also temporary, just like so many real-life relationships can be: after keeping things going long-distance for a little while, the two decided they were better off as just friends.

And that's where things might have ended for our Rec Room friends. In fact, it *was* where things ended by the time I'd finished the first draft of this book. But it turned out that something a little more significant was in the making. (All together now: *cut to me, hunched over my laptop in despair.*)

The "crush" Priscilla was referring to? That was also a Rec Room user—a guy named Mark. Mark's been self-employed for the past dozen years, running a collection of search-based websites. After his mother retired recently, he helped her relocate to a small town outside Seattle. The town was great for seniors, but it wasn't so great for thirty-somethings—especially thirty-somethings who work from home. Cue Rec Room. "This is pretty much my outlet," he tells me the first time we meet in VR, "unless I want to drive two and a half hours to Seattle and then two and a half hours back just to go to a club." Plus, it's exercise: two or three hours a day ducking and jumping in VR paintball beats getting on the rowing machine.

Mark was part of the same extended group of friends as Priscilla and Ben. On weekend nights, they'd all hang out, drink, play games, and tell embarrassing stories about themselves—"like you're ten years younger than you actually are," Mark jokes. He's easygoing but has enough gravitas to counterbalance his silliness; it's an appealing combination. Priscilla developed some feelings for Mark, but then her friendship with Ben blossomed.

Months later—long after the Ben saga—I got a text from Priscilla. Well, a series of texts.

Hey

So I married someone from Rec room

And it wasn't blitz

That someone, in case you hadn't guessed, was Mark. After some time away from Rec Room, he had come around again, and over the course of the next few weeks his and Priscilla's conversations took on a different tenor than they had before. They recognized themselves in each other; they confided in one another. Finally, unable to wait any longer, they decided to meet, and Mark flew to Alabama. Just in case, he brought an engagement ring with him. That was a Tuesday; on Thursday, they were engaged; on Friday, they were married in a gazebo high on a hill outside Birmingham. They broke the news to their Rec Room friends on Discord; the following week, with Mark back in Washington, they held a follow-up ceremony (and drunken reception) in VR.

The Rec Room wedding was everything you'd imagine about a ceremony populated by cartoony avatars. The platform had recently introduced a "maker pen" that let users create 3-D objects, and it was put to great use: a virtual bouquet of flowers; a pink, three-tiered virtual cake. Priscilla wore a pale yellow wrap dress and a garland of purple daisies, Mark a dark suit and a top hat, red cuffs peeking out from under his suit jacket. The bridesmaids were in blue, the groomsmen in tuxedos, and they all gathered in a gazebo at dusk. Later, they danced—each of them in their own headsets, standing in front of their computers, 2,600 miles apart.

Ben was there, too. After all, when you've spent that much time with another person, and your friendship goes that deep, you go to their wedding. Besides, it's not like you're shelling out for a plane ticket.

This isn't the first time people have started a relationship because of VR; this isn't even the first time people have gotten married in VR. (The first time was back in 1994, when a San Francisco couple got married at Cybermind, a VR arcade where the bride worked.) Altspace hosted a VR wedding last year. But in both of those examples, the couple getting married had met in real life; they just tied the knot in VR because it was something new. Priscilla and Mark, though? That just might be the first time people got married in VR after meeting in VR.

It won't be the last, though. Not as long as our hands and heads are in there, helping our hearts connect with other people as we never have before.

# 8

# REACH OUT AND
# TOUCH SOMEONE

## HAPTICS, TACTILE PRESENCE,
## AND MAKING VR PHYSICAL

IF YOU'RE LOOKING for the forefront of VR as entertainment, you'd likely guess it's somewhere in Los Angeles or New York, where densely clustered content companies are pioneering new forms of storytelling. Maybe you'd take a flyer on the Bay Area, where billion-dollar companies are sinking millions into research. Chances are you wouldn't say, "Oh, about a half hour south of Salt Lake City." But that's where I am—in Lindon, Utah, to be precise, on a pancake-flat plateau of land sandwiched between Utah Lake and the mountains of the Wasatch Range.

This whole area, the stretch of Interstate 15 from Salt Lake City down to Provo, has dubbed itself Silicon Slopes in solidarity with so many other tech-heavy areas. Ancestry.com and Overstock both have headquarters nearby. So does DigiCert, which—to make a technical story short—helps web browsers verify the authenticity of secure websites. In fact, DigiCert's founder is the reason for my visit.

Ken Bretschneider grew up in an Ontario fishing town that took its holidays seriously, and Halloween was his favorite; when Ken was a kid, his friend's father transformed their garage into a maze of horrors. In 2008, a grown-up Bretschneider turned his own Utah home into a haunted house—and then did so each year thereafter, ultimately expanding to occupy a half acre of land and bringing in more than ten thousand people. He decided to turn the seasonal idea into a year-round venture, and in 2014 he unveiled his plan: Evermore Park, as he called it, would be a Victorian-era "adventure park" that prioritized immersion over thrills. Period-appropriate employees would interact with visitors, and the park would change to adapt to holidays like Halloween and Christmas. Think Disneyland mixed with Colonial Williamsburg. (Actually, think Disneyland mixed with *Brooklyn* Williamsburg—the mustaches are a better match.)

But at the same time Evermore was getting planned, another idea was taking shape. Two of the people Bretschneider had hired to help make Evermore a reality were Curtis Hickman, a professional magician and illusion builder, and James Jensen, a multimedia designer. Jensen would create a digital version of the park as it took shape. Seeing the digitally rendered Evermore made them think of VR—and Jensen pitched an idea he'd wanted to realize since the early 2000s. He had been working on a scuttled movie adaptation of Little Red Riding Hood; it leaned heavily on computer-generated backgrounds and used a position tracker on the camera to help directors see where actors should stand for proper placement in the CGI world. *Why don't we try to map a VR world over a real physical space?* Jensen asked Bretschneider and Hickman.

That idea became The VOID.

## ENTER THE VOID

The VOID—in suitably grandiose fashion, it stands for Vision of Infinite Dimensions—is the realization of Jensen's dream. It's part of a wave of "location-based" VR facilities that are merging presence with sport for activities that are half holodeck, half laser tag. The experience hinges on creating a VR video game that's physically re-created on a large soundstage. Any feature you see in the game is also in the real world, so if in VR you see a wall, and you reach out your hand to touch that wall, your flesh-and-blood hand will touch an actual wall. The soundstage is ringed by a huge array of trackers, which allows you to roam over a much larger distance than you'd be able to with a home VR setup. See a chair in your headset? Walk over to it and sit down—like, bend your knees and lower your butt—and you'll find your corporeal hiney supported by a chair that's the same size as the one you see.

Two adventures are up and running when I visit: an Indiana Jones–style jungle adventure called "The Curse of the Serpent's Eye," and a second experience set in the *Ghostbusters* universe, and blessed by Ivan Reitman himself, the film's director. Both can accommodate up to four players at a time, but I'm going to go through each of them with only one other person.

Before that happens, though, I need the proper gear. That's why I'm standing here in The VOID's prep area, re-creating one of my all-time favorite movie tropes: the "suiting up" montage that's familiar to any action-movie fan. First, I shrug on a heavy vest they call the backtop. It's got big plastic patches on the front that rumble and buzz, to give me sensory feedback, and a brawny laptop on the back, along with a big battery and some more sensory modules. The headset connects to the laptop, which turns me into a self-contained unit, able to move around the soundstage freely.

Next up: the headset, which is basically a souped-up version of an Oculus Rift. It's got *huge* headphones, and slides down over my face like a sci-fi visor. Small silver balls that ring the visor ensure that the tracking system high overhead can follow me accurately no matter what crazy contortions I make on the soundstage. And perhaps most noteworthy, a small Leap Motion module attached to the front of the visor brings my hands into VR, no controllers necessary. With my headset on, I hold my hand out in front of me—and there's a hand, fingers wiggling just like mine.

Enough standing around. There's a door in front of me with a hand symbol on it. I push it open and walk through, the feedback modules on my chest and back buzzing, and find myself walking down a stone hallway in the ruins of a Mayan temple. And when I say I'm walking, I don't mean a couple of steps—I mean I'm *walking*—like, more than I should be able to on a thirty-foot-by-thirty-foot stage. That's because of "redirected walking," a VR technique that takes advantage of humans' actually-kinda-horrible navigation skills.

The thing is, you stink at walking in a straight line. Yes, you. And me, and everyone else as well. As a rule, people go off course when they can't see where they're going; that's why we began using the stars as a guide when traveling at night or on open water. Redirected walking takes advantage of that fuzzy sense of direction by presenting you with a path in your headset that's slightly different from the one you're actually walking on. Your vestibular system is far more finicky than your vision; as long as your eyes tell you you're moving in the same general direction that your inner ear feels, everything's fine. So in your headset, you might think you see a straight hallway ahead of you, but you're actually walking in a curved line—and a VR experience can turn you more than 49 percent more than you *think* you've rotated without your noticing. At The VOID, redirected walking is how you can roam through an expansive adventure like "The Curse of the Serpent's Eye" while in reality tracing a circuitous, but compact, path on a surprisingly small soundstage.

There's more to my adventure than just walking, though. There are "4-D effects," as the theme-park world has come to call them: misters and fans that blow cool air and moisture across my face to simulate the jungle environs, motors in floor panels that can make me feel like the temple is collapsing. There are objects that belie their pedestrian real-life appearance in VR: what looks like a spray-painted club with some tracking balls on it becomes a torch I can pick up and light on a brazier. Likewise, that brazier is just a radiant heater behind a grate, but in VR I can still feel its warmth on my hands. When I hold the lit torch up to a seal on a door and the door explodes, I can feel the concussive impact in my chest—and reach out to steady myself against a wall. These sorts of immersive touches simply aren't possible in a home setup, but that's the whole point of location-based VR. It's presence on a grander scale.

That presence is also compounded by the fact that in The VOID, I'm not alone. In "The Curse of the Serpent's Eye," I can hand the torch to my companion, and because she sees it in her headset, the hand-off is perfect. Sure, I can hand an "object" to another "person" in regular VR, but every part of that is simulation. The object is my hand controller with the trigger held down; when I let go of the trigger and thus the "object," I've still got the controller in my hand. The other person might be another person *somewhere,* but in VR she's just an avatar—our fingers can't brush against each other the way they do in The VOID.

And that's just an exploratory adventure. The *Ghostbusters* experience adds action to the equation. Along with the backtop and headset, my companion and I tote big rifles; in VR, those backpacks and rifles turn into proton packs and blasters, because *we're* Ghostbusters.

Let me repeat that: WE'RE GHOSTBUSTERS.

Look, I was ten when that movie came out, and in the thirty-odd years since, I've seen it dozens more times. I've made more than one "last of the Meketrex supplicants" joke at work. I truly believe it's one of the greatest comedies—nay, movies—ever made. So maybe I'm not the most objective source here. But I've gone through more VR experiences than I can count, from gaming to entertainment to social to spiritual to sexual (don't worry, we'll get there), and I'm here to tell you that standing on a rickety metal catwalk on the edge of a building, its chain railing bouncing against my thighs, yelling at my co-worker to cross the streams so we can take out the Stay Puft Marshmallow Man, might just be the most fun I've ever had in there.

But why? What do you *call* this kind of presence? Well, to start to answer that, we're going to need to go from a virtual Mayan temple to a different kind of Temple.

## COPRESENCE

In 2003, Shanyang Zhao, a sociologist at Temple University, published a paper called "Toward a Taxonomy of Copresence." It's very smart and very long and not strictly about virtual reality, but its chief purpose is to establish a system that accounts for the various ways that two people can be together. (Or, in his terms, "the conditions in which human individuals interact with one another face to face from body to body.") This was especially important, he wrote, to account for the ways in which the internet had expanded the parameters of what we mean when we talk of being "with" someone.

Zhao laid out two different criteria. The first was whether or not two people are actually in the same place—basically, are they or their stand-ins physically close enough to be able to communicate without any other tools? Two people, he wrote, can either have "physical proximity" or "electronic proximity," the latter being some sort of networked connection. The second criterion was whether each person is *corporeally* there; in other words, is it their actual flesh-and-blood body? This second condition can have three outcomes: *both* people can be there corporeally; *neither* can be there corporeally, instead using some sort of stand-in like an avatar or a robot; or just one of them can be there corporeally, with the other using a stand-in.

The various combinations of those two criteria lead to six different types of copresence, Zhao wrote. "Corporeal copresence" is plain old face-to-face physical contact: two people in a coffee shop. If those real people are networked together and can communicate face-to-face, like a Skype call, that's "corporeal telecopresence." And if something isn't corporeal, it's virtual. "Virtual copresence" is when a flesh-and-blood person interacts physically with a representative of a human; if that sounds confusing, a good example is using an ATM, where the ATM is a stand-in for a bank teller. "Virtual telecopresence" replaces a physically present thing like an ATM—something that spits out money when you insert a card and press some buttons—and replaces it with, for instance, Waze's turn-by-turn navigation, which you listen to while you drive but is "tele" because it's being delivered from a network server.

171

Got it so far? Don't worry, there's no quiz. It does get a little more difficult, though. See, there's "hypervirtual copresence," which involves nonhuman devices that are interacting in the same physical space in a humanlike fashion. This is pretty uncommon, but Zhao uses the example of robots playing soccer. And finally, "hypervirtual telecopresence," in which nonhuman stand-ins interact via networking—like two bots communicating over the internet.

So that's copresence. Except something is missing. That something is social VR. It's obviously copresence, but it doesn't quite fit into any of these categories. Zhao refers to this sort of hybrid as a "synthetic environment" and claims that it's a combination of corporeal telecopresence (like Skyping) and virtual telecopresence (like Waze directions)—"human individuals [interacting] with each other remotely in real time via avatars that operate in virtual settings."

Checks out, right? That sounds like every form of social VR we've talked about so far. Well, every form except something like The VOID. It's a "synthetic environment," sure, but one that takes its cues from a corporeal environment—and vice versa. It's physical proximity *and* electronic proximity, blended together to create an entirely new type of immersion.

So new, in fact, that it doesn't exactly have a name yet. For now, let's call it *tactile presence*.

## HAPTIC TACTICS FOR TACTILE APTNESS

Of the five human senses, a VR headset can currently stimulate only two: vision and hearing. That leaves three others—and while smell and taste may come someday, for now let's just file those away under Slightly Creepy Gimmick. (That doesn't mean people aren't working on both of those, though. Some researchers have shown off an "olfactory display" made out of micropumps and acoustic devices, and others have worked out how to electrostimulate the tongue to induce taste sensations. Both of those projects, it may not surprise you to learn, were developed in Japan.) What really matters for VR right now, and in the coming decades, is touch. Touch is how we comfort each other; it's how we please each other. Even *almost*-touch can be a factor, because the heat we all radiate arouses tactile sensations. Remember how being excluded in a VR game of catch could make people act more antisocial in real life? Well, being caressed with slow, "affective" touches can mitigate the feelings of ostracization that can arise from that same VR game.

Communicating that touch across distance, though, might just be the most difficult piece of the presence puzzle. Imagine wearing a VR headset and reaching out your hand to touch a surface. The VOID's magic is that it presents a solid surface for your hand to touch—but what if you were at home? And what if it weren't a wall you were reaching out for, but another person? Even more difficult, what if that person were reaching out *their* hand to touch yours? How will we get to a point where that becomes not just possible, but realistic?

I don't know, to be completely honest. But given what we're capable of doing now, and how we've gotten here, and what people are experimenting with, we at least have some sense of a road map. So let's start by talking about *haptics*. The word just means "relating to the sense of touch," but it has become an increasingly important field in the world of what's known as "human-computer interaction," and now it's often used to refer to technology that seeks to re-create touch.

The idea of haptic feedback to create tactile presence has been around since at least 1932. That's when Aldous Huxley's future-set novel *Brave New World* imagined the "feelies"—movies that induced physical sensations in viewers that matched up with what was happening on the screen. In Huxley's mind, this was a satirical extension of the "talkies" that had evolved from silent movies a few years earlier, but the book's memorable (and, okay, straight-up offensive) feely scene offered up a use that was more than just satire:

> The house lights went down. . . . "Take hold of those metal knobs on the arms of your chair," whispered Lenina. "Otherwise you won't get any of the feely effects."
>
> The Savage did as he was told.
>
> Those fiery letters, meanwhile, had disappeared; there were ten seconds of complete darkness; then suddenly, dazzling and incomparably more solid-looking than they would have seemed in actual flesh and blood, far more real than reality, there stood the stereoscopic images, locked in one another's arms, of a gigantic negro and a golden-haired young brachycephalic Beta-Plus female.
>
> The Savage started. That sensation on his lips! He lifted a hand to his mouth; the titillation ceased; let his hand fall back on the metal knob; it began again. The scent organ, meanwhile, breathed pure musk. Expiringly, a sound-track super-dove cooed "Oo-ooh"; and vibrating only thirty-two times a second, a deeper than African bass made answer: "Aa-aah." "Ooh-ah! Ooh-ah!" the stereoscopic lips came together again, and once more the

facial erogenous zones of the six thousand spectators in the Alhambra tingled with almost intolerable galvanic pleasure. "Ooh . . ."

That idea of induced tactile pleasure would pop up in sci-fi books and movies again and again, from the Orgasmatron in Woody Allen's 1973 spoof *Sleeper* to the electrode-link VR hookup in *Demolition Man.* (I know I've mentioned it before, but *it's just so weird.*) Given that personal computers were still decades away, though, the only haptic feedback that really entered the public consciousness came courtesy of "Magic Fingers," which made your motel bed vibrate for a quarter. Novelty aside, haptic interaction was largely confined to mechanical devices, like those in the 1940s that would let workers remotely handle hazardous materials.

Moving from these analog examples to digital ones began to happen in the 1960s in research labs, and in 1971 one of those researchers published a doctoral dissertation that detailed a surprising breakthrough. "A computer helping an individual feel some object which existed only in the memory of the computer could justifiably seem to be a far-out idea," wrote A. Michael Noll—but he had devised a machine which could accomplish just that.

His invention looked like a metal cube filled with a ceiling-mounted array of beams and motors. On its top was a joystick-like device that could be moved in all three dimensions. The beams and motors could make the stick more difficult to maneuver, according to instructions it received from a computer. If users moved the stick around enough, taking note of where it stopped, they could deduce that they were moving the device along the outside of an invisible cube, or the inside of a sphere.

Noll described it as "similar to a blind person exploring and poking around three-dimensional shapes and objects with the tip of a hand-held pencil." It was the first time such a thing was possible, and Noll seemed to sense that even something so "far-out" had wide-ranging applications. He imagined a person in New York City "feeling" a cloth produced by a manufacturer in Tokyo. "'Teleportation' in one sense would be closer to reality," he wrote.

Now that he had invented a way for people to feel virtual objects, Noll planned to build a 3-D head-mounted display that could show the computer shape—essentially the first VR headset. However, he accepted a job with the office of President Richard Nixon's science advisor and never returned to that research.

More than forty years later, Noll seemed to be disappointed with the amount of progress VR had made without him. "It is perplexing that with the advances in technology that have occurred since the early 1970s that what is today called 'virtual reality' and 'haptic' seem behind our vision back then," he wrote in 2016. "I therefore issue my challenge to today's community to create what was envisioned decades ago. Otherwise, much of today's virtual reality indeed is little more than real fantasy."

Noll's work was groundbreaking. But where did the rest of the world get its first glimpse of synchronized haptic feedback? Video games, of course. In 1976, still the early days of arcades, Sega released a game called *Moto-Cross* (later to be rebranded as *Fonz* in a bald attempt to capitalize on the *Happy Days* character's popularity) in which players controlled their tiny motorcycle-riding characters via a set of handlebars mounted to the game. If the bike collided with another, those handlebars would vibrate in players' hands, making them feel the "collision."

The technology was simple, essentially a very localized version of Magic Fingers, but it opened up a new world of so-called force feedback in video games. Haptic features spread to driving games, then to other arcade machines—and in the meantime, to pagers and cellular phones—and in 1997, they entered the home. Nintendo began selling Rumble Pak, a small module that attached to the bottom of one of the company's game controllers for use with now-classic games like *Star Fox* or *GoldenEye 007;* when players steered their ships into others or fired guns, the controller vibrated appropriately. Thereafter, virtually every video game console came with controllers that had force feedback built into them.

When VR rearrived, haptics became significantly more important. The added immersion that came from a vibrating steering wheel, or the ability to feel recoil in a shooting game, paled in comparison to hand presence. Looking at your "hands" in VR, the controllers you're holding melt away. Your brain effectively collapses the boundary between flesh and artifice, so haptic feedback is no longer a matter of feeling the controller you're holding, but of holding the feeling.

The hand presence that devices like the Oculus Touch enable, though, are merely a first step toward *tactile* presence, truly immersive sensations of touch. Like conventional video game controllers before them, today's VR controllers use tiny motors embedded within them. These motors and their vibrations can communicate impacts of various strength and duration—buzzes, taps, knocks, thuds—but little else. There's no shape, no weight, no texture. That's the magic of The VOID, and other VR facilities like it: they induce tactile presence through real-world objects. But in order for tactile presence to get out of The VOID and into the Metaverse, for Noll's vision of sampling suit cloth from a continent away to go from far-out to attainable, we're going to need to be able to experience those properties. No one way gives us all of those things yet, but a number of solutions exist that can provide one or two pieces of the puzzle.

**Haptic-feedback accessories** take the premise of force feedback—using rumble to communicate impact and contact—and distribute the feedback over more of your body, rather than just your hands. At The VOID, this takes the form of a vest packed with vibrating patches that go off in sync with explosions. However, that's just the beginning. Lotte World, a popular indoor attraction in Seoul, South Korea, features a VR experience that puts each user in a five-hundred-dollar suit packed with eighty-seven distinct feedback points; that's enough to make it feel as if a zombie is raking its claws across your back.

**Gloves** have been a part of the public's fascination with VR since the Nintendo Power Glove, a short-lived late-'80s video game peripheral that allowed people to control a game with their gestures. (Although the Power Glove wasn't a VR device, it was the brainchild of the same pioneers who had come up with the Data Glove, which *was*.) By now, you know that despite Mark Zuckerberg's Spider-Man gesture in Chapter 3, gloves aren't necessary as an input device; thanks to outward-facing sensors like Leap Motion that can track our individual fingers, we can use our own hands. Gloves do, though, have a distinct advantage in output: because they cover your hands completely, they're able to provide much more detailed haptic feedback. This means that if you're using, say, a virtual keyboard, you're able to feel a little tap or buzz on the relevant finger when you "press" a key—intensifying the sensation and likely increasing your speed and accuracy.

**Full body suits** are the logical end of direct haptic feedback.

Like gloves, they've been in the cultural imagination for decades, largely as input devices. (Once again, I direct you to *Lawnmower Man*, this time for the wetsuits that allowed people to move freely in VR.) However, like gloves, the output potential is far more interesting for presence, and especially for intimacy. A haptics company called HaptX has secured a patent for what it calls a "whole-body human-computer interface" that's basically a VR nerd's ultimate dream. Imagine a wetsuit, its interior lined with tiny actuators that can deliver temperature and pressure changes. Now imagine that you put *that* on and then climb into an exoskeleton that suspends you in the air and allows you to move freely, while still delivering actual force feedback—like, making it more or less difficult to move various body parts. That's more than "rumble"; that's damn near groundbreaking.

Or it would be, if it ever panned out. Is HaptX anywhere near having a consumer product? Kind of. The company has announced plans to release a glove in 2018, and even brought it to Sundance this year. (Previously, HaptX would show off its technology via a box you stuck your hand into.) In previous demos, a small deer walks across your hand—and you can feel its legs skittering on your palm. The demo chills your hand when a virtual snowball is placed on it and warms it when a virtual dragon breathes fire in your direction. That "whole-body human-computer interface" isn't something we're talking about for tomorrow, or next year, or even longer—but as proof of concept, it ain't bad.

These are all wearable options, but there are a lot of other ones as well that might just end up helping transport your body along with your mind—like, say, using **ultrasonics,** low-frequency sound waves, to approximate objects that your hands can feel. I know it sounds odd, but it works. Not long ago, I stood with a headset on my face and my hand hovering over a small square board. The board was lined with a dense, orderly array of small black circles, each of them an "ultrasonic transducer"—essentially a tiny speaker capable of pumping out bursts of acoustic pressure.

In the headset, I saw a cube and a ball; when I reached my hand out to push them around, I could *feel* them. They didn't have heft, and they didn't exactly feel solid—if I tried to cup them in my palm, they disappeared, since the ultrasound waves couldn't pass through my hand—but I could sense their presence, and even distinguish the edges of the cube and the roundness of the ball.

VR may not even be the place we see (well, feel) ultrasonic haptics in our lives for the first time. The company that created the VR demo I experienced, Ultrahaptics, has sold development kits to a number of car companies, some of which are doubtless experimenting with ways to create invisible dials and knobs for the instrument panel of the future. But even if VR is second or third in line, ultrasonics' applications may go far beyond holding objects. In fact, because of its formlessness, it might be the haptic technology best suited to approximate the effects of human touch.

I'm not just saying that idly; a collaboration between an interdisciplinary team of British academic researchers and the founder of Ultrahaptics (who came up with the idea as a graduate student at the University of Bristol) found that the technology could be used to communicate emotion through the air. In their study, they asked a group of ten people to use the Ultrahaptics device to create patterns based on images like a car on fire, a graveyard, and a calm scene with trees. A second group of ten people then felt all the patterns and narrowed them down to what they considered to be the best "touch" for each image. Last, yet *another* group of ten people felt the patterns that made the cut, except the patterns were presented in a random order that didn't necessarily match with the image the pattern had been created for—and, when asked to rate how well the "touch" matched with the images, they significantly preferred the pairings that had actually been created by the first group. It was like a game of telephone; the third group had no way to know what the first group had done, yet as a whole the entire group of subjects had settled on touch patterns that they agreed fit with the moods of the images.

"Our findings suggest that for a positive emotion through haptic stimulation one might want to stimulate the area around the thumb, the index finger and the middle part of the palm," the researchers wrote. Similarly, "if one wants to elicit negative emotions . . . the area around the little finger (pinky) and the outer parts of the palm become a relevant design space." The article outlines not just *areas* of the hand that might elicit certain emotional reactions, but also the *direction* of the touch (vertical simulation toward a person is positive, while caresses moving away from the palm and toward the fingers is negative) and even the sonic frequency and duration of the touch. This is just a starting point, to be sure, but it suggests a path that may help us unlock, and even codify, the power of touch in VR.

But what about texture? How might we be able to distinguish soft pima cotton from pebbled leather—or even smooth skin? As difficult as communicating the shape of an object is, texture is even more difficult. Yet, a pair of prototype controllers that Microsoft's research department has built might be able to do both of these things. One of them, called NormalTouch, is in many ways a tiny, handheld version of A. Michael Noll's 1971 breakthrough haptic device: a small pad underneath the fingertip can tilt and telescope in keeping with the contours of a virtual object. Its companion,

TextureTouch, adds to that fingertip pad sixteen tiny pillars that can extend or retract in order to help you feel, say, the ridges on a statue. It may also let you feel some other things; as one commenter on a YouTube video demonstrating the controllers asked, "can i touch virtual tity's." (Thanks, Smack Thrustcrusher, whoever you are; your spelling, grammar, and punctuation skills are a credit to keyboard-bound horndogs everywhere.)

Granted, the fact that Microsoft is exploring this stuff is about as surprising as Bill Gates wearing a sweater—but you can't say the same for one of the other major companies that might unlock the path to feeling texture in VR. Disney's research department has been investigating haptics for years. Some of that is because it's an integral part of the theme-park experience, but the company also clearly has an eye on the future; scientists working in Disney Research's Pittsburgh headquarters have released multiple papers detailing tactile feedback technology for both VR and its cousin, augmented reality. (We'll get a bit more into AR later on.) One of its most interesting projects involves creating an electrical field around a person's fingers, which can then be manipulated, allowing programmers to make a smooth real-world object feel to the user as though it's bumpy. "In a broad sense," the researchers wrote in an article detailing the project, "we are programmatically controlling the user's tactile perception."

All of which is to say: the human brain, as remarkable as it is, is hackable. So yes, tactile presence of the type we experience in The VOID—solid, textured, immovable objects we can interact with—is currently exclusively an IRL-only phenomenon. But the idea of inducing such sensations virtually isn't just plausible. Increasingly, it's *probable*.

## LOCATION, LOCATION, LOCATION: YOUR FIRST MIND-BLOWING VR EXPERIENCE MIGHT NOT BE AT HOME

Meanwhile, facilities like The VOID have become one of the most vibrant of VR's many dimensions. Whether you think of them as arcades or theme parks, they offer a number of distinct advantages over a home setup. For one, depending on the gear, you can roam freely inside a much larger space than an off-the-shelf VR system can manage. Thanks to the wonders of redirected walking, it doesn't take a football field to make you think you're on a football field. Besides, these venues are sinking some serious cash into a bespoke setup, from ultra-high-powered PCs to tracking systems to headsets that aren't even available for home use.

One of those, StarVR, looks like something out of an episode of *Star Trek*. It's not uncomfortable, but it's *huge*, with a display that extends around the sides of your head. Inside the headset, the virtual world goes well past the edges of your peripheral vision. This is the headset that IMAX uses at its four dedicated VR centers—in Los Angeles, New York City, Shanghai, and Toronto—as well as in an enormous VR center that opened in Dubai at the end of 2017.

And one of the games you can play on that StarVR headset, at that Dubai arcade no less, perfectly illustrates why all this talk about haptics and tactile presence matters. (And before you ask: no, I didn't get to go to Dubai. I played it in a conference room in a San Francisco hotel during a VR event. Then again, it was in the middle of a crazy heat wave, and the hotel's air conditioning was clearly outmatched, so it wasn't *entirely* unlike Dubai.)

When I walked into the room, all I knew was I'd be playing something called *Ape-X*, but I had no idea what it entailed—and I certainly didn't anticipate the setup. On the ground was a hexagonal metal grate, maybe five feet across; a metal pillar rose out of the center of the grate. Before I knew it, one of the developers was helping me situate the giant headset over my eyes. Then the headphones went on. Then I saw two massive metal gauntlets floating in my headset, almost like metal Hulk hands. These were real controllers, as it turned out; when I reached out my hands, the developer helped slide them on.

Finally, the game started, and I understood what the grate and pillar were for. I was standing on a narrow circular catwalk ringing the uppermost spire of a skyscraper. I peeked over the edge and saw a line of cars a hundred feet below—but the cars were *flying* and were themselves hundreds of feet above the ground. I was impossibly high up, perched precariously in a situation that I legitimately have had nightmares about. Even crazier, I was clearly the titular Ape X (at the apex! Yeah, yeah, we all get the title), an escaped super-intelligent simian, a sci-fi update of King Kong. I had no Fay Wray, just my enormous gauntlets—which, thankfully, were armed with laser rifles and a small supply of guided missiles.

I'd need them all, too; for the next seven minutes, I had to fend off wave after wave of flying enemies. Some I could shoot out of the sky. Others came so close I had to knock them away, swinging my huge gauntlets. Still others came flying in from directions where I wasn't looking, strafing me with their own laser fire and forcing me to shuffle around the catwalk, using the skyscraper's spire as cover. When I wasn't able to keep an arm hooked around it, I pushed my back against the spire as best I could, both for defense and for some semblance of stability in the chaos I was trying to navigate.

My lasers felled the last enemy. Finally, I thought, the game might end. And it did . . . but not before one more unexpected challenge. A hovercraft pulled up next to the catwalk, like a tiny floating metal barge. It had been sent to help me escape, but I'd have to board it, which meant not only peeling myself off the spire that had become my one friend in the world, but actually taking a step off the grated catwalk. *Do it,* my rational brain said to me. *You know you're in VR. You know that out there, the grate doesn't end in nothingness—it ends in carpet.* My rational brain may know what it knows, but my reptile brain only knows what it *feels*. And at that moment, my reptile brain was feeling like I was hundreds of feet in the air and my knees might never unlock.

So there I stood, the two sides of my brain deliberating the nature of presence for what felt like *another* seven minutes, until the rational half willed my leg to move. Slowly, I stepped one foot toward the hovercraft, my other leg bending to jump. And I jumped . . . directly into the waiting arms of the developer, who was making sure I didn't go ass over teakettle into the expensive computer setup. "Nice work," he said amiably. "Not a lot of people end up jumping."

And honestly? If I were back in there right now, I might not be able to jump again. With the grate under my feet and the spire's comforting solidity at my back, I was more there than I was anywhere else, rational brain or no. The things we can't yet sense in VR—true tactile presence, with heft and texture and warmth—are the very things that intensify the feelings we *do* have there. The question remains, though, what that tactile presence might mean for the connections we form.

For that, we're going to need to go on a blind date.

## THE DATING GAME:
## HOW TOUCH CHANGES INTIMACY

John and Shelby met fewer than five minutes ago, but they're already slow-dancing outside in the night air.

"Okay, I have a question for you," Shelby says, her blonde hair curiously still as she turns her head. "Do you dance well because you're a Southerner?"

"Actually, I grew up all over the country, so I don't really consider myself a Southerner," John says.

"Are you an army brat?"

"No, I grew up *poor*," he says in a sing-songy voice, letting go of her waist and extending himself outward like Fred Astaire. "So we went *all* over the country to find *different* homes and *different* places to live." His tone might be a little glib, given what he's telling her, but it's hard not to marvel at his candor, especially with someone who's effectively a complete stranger. And whose feet are on backwards. And who doesn't seem to mind that he's hovering half a foot above the ground. Under the light of the Earth, it seems, anything is possible.

*Wait, under the light of the what?*

Yeah, you know this move by now, right?

*Ha ha, clearly these are avatars and they're dancing in VR, and what, now you're going to talk about presence and intimacy and being on the moon?*

Mostly, sure. But something else is going on here too. Because while John and Shelby are dancing on the moon, and eventually turning into aliens and astronauts and *T. rexes* and cardboard box people and cacti and skeletons and otherwise enjoying the novelty of imagination becoming flesh, their dance isn't as clumsy and claylike as it looks. ("I like the way my hand goes through your entire neck," John says to Shelby at one point.) In VR, their physical interactions are sub–*Gong Show* levels of terrible, but in the studio where they're wearing headsets and motion-capture suits, they're actually not *un*-graceful.

See, there's meeting someone in real life, and then there's meeting someone in VR. But what do you call it when it's meeting someone in both at the same time? When tactile presence gets involved, and you can blend the virtual and the real not just in your eyes and ears and brain, but on your *skin*, how does that change the intensity of an experience? And, more important, how does that change the way a relationship evolves?

Let's back up a bit. Like twenty-five years.

## PRESENCE TENSE

In 1992, MIT Press began publishing *Presence*, the first academic journal that sought to unify an interdisciplinary exploration of what it called "teleoperator and virtual environment systems." People with all kinds of backgrounds contributed: engineering, computer science, media, and the arts. (Fun fact: one of the articles in the very first issue was by Warren Robinett, whose name might be familiar to gamers as the person who created what's widely considered the first "easter egg" in a video game.) And even then, in the early days of civilian VR, those people agreed on a general definition of presence. But people were just starting to figure out how to classify, let alone *measure*, such a slippery concept.

Some of the first to try were Bob Witmer and Michael Singer, two army researchers who had cooked up a thirty-two-item questionnaire for researchers to use in experiments. By asking volunteers to answer questions like "How natural did your interactions with the environment seem?" and "How distracting was the control mechanism?," they hoped they would be able to begin to delineate the many factors that contribute to presence. After using that questionnaire in four experiments, Witmer and Singer broke presence down into four categories—users' perceived control, sensory stimulation, distraction, and realism—and further identified a total of seventeen subcategories that constituted the building blocks of the phenomenon. "We do not claim to have identified all of the factors that affect presence," they wrote in their conclusions, "nor do we fully understand the presence construct, but we believe we have made considerable progress."

That questionnaire became widely cited in the field of presence research—as much for the responses it engendered from other researchers as for its influence. Not only did another questionnaire arise just about simultaneously with Witmer and Singer's, but its authors used these questionnaires to show that they were an intrinsically flawed way to discuss presence: the questionnaires, they argued, couldn't even distinguish between a real experience and a virtual one. (Granted, that might be the platonic ideal of VR, but in the late 1990s, when this war was happening, that "virtual experience" looked like something someone created in fifteen minutes using Microsoft Paint.)

This sort of discussion went on for years and bears mentioning here only to point out that its utility didn't really go outside the psych lab. Methodological explorations of presence may have clinical applications, but they were basically the province of the science community. However, as time went on, people started thinking about presence more holistically, taking it out of the lab and exploring it from the user's perspective. Then in 2010, one study mashed up the world of video games and presence questionnaires to cook up an entirely new way of thinking about presence.

Three researchers—two professors at the University of Central Florida and one at a private research company in Maryland—took their inspiration not from the world of VR, but from the world of design. In recent years, the concept of "experiential design" had emerged as a bit of a buzzword, blending together disparate fields like psychology, brand strategy, and theater to create a multidisciplinary approach to design. The researchers used the underlying principles of experiential design to create a questionnaire that hewed to a new taxonomy of presence—five categories that they saw coming together to create presence for a user.

**Sensory:** The stimuli created by the hardware—visual display or haptic feedback

**Cognitive:** Mental engagement, like solving mysteries

**Affective:** The ability of a virtual environment to provoke a fitting emotional response

**Active:** Empathy or other personal connection to the virtual world

**Relational:** The social aspects of an experience

They then used the video game *Mirror's Edge* to test their new system. (If you don't remember that one, it was a first-person action game in which you race across the tops of skyscrapers; think of it as parkour on meth.) Despite the game not even being

189

in VR—this was just a plain old Xbox 360 game played on a plain old television that wasn't even high-def—the researchers' hypotheses held, meaning that there was at least some merit to thinking about presence this way.

This is only one proposed classification of presence, of course; there are others as well. But clearly, it resonates with everything we've been talking about so far—especially the affective, active, and relational aspects. There's one thing that's missing from this system, though, and that's the effect of tactile presence.

## GLITCHING TOWARD BETHLEHEM

Ryan Staake has creativity in his genes. His father, Bob, is an accomplished illustrator who is responsible for some of the most iconic *New Yorker* covers in recent memory (a onetime St. Louis resident, Bob responded to the 2014 unrest in Ferguson, Missouri, by depicting the Gateway Arch as half black and half white, with an unbridgeable gap in the middle). So the fact that Ryan has spent his young career evolving should come as no surprise. A graphic designer by trade, he started out working on user interfaces at Apple and then founded a production studio and moved into making music videos. Some of them contained the seeds of VR, like 360-degree video or digitized versions of musical artists.

But if you know Staake's work, it's probably because of a fortuitous failure. One video he directed, an elaborate production for a song by Young Thug, fell apart when the rapper never showed up to the shoot—so instead Staake compiled all the footage he had shot, interspersed with title cards explaining what was *supposed* to happen. The result went viral, pulling in nearly thirty million views on YouTube and prompting coverage from seemingly every music publication on the internet.

In the aftermath, a screening series in LA invited Staake to come out from New York for an onstage interview; he couldn't go, but he did the next best thing. Using a motion-capture suit and a scanning system that he and his colleague had cooked up, he recorded a VR video in which a digitized version of himself pretended to take questions from an imaginary live audience.

The result was hilariously imperfect: his avatar's right hand seemed to have its middle finger permanently extended, and his iPad-scanned face with its unmoving mouth could easily have been the mayor of Uncanny Valley. His legs bowed out cartoonishly; his arms bent at the elbows like they were paper towel rolls. At the end, avatar-Staake walked over to a pristine digital sports car and got in—only to have his arm stick out through the middle of the closed door. It was a self-aware mess, and was all the more charming for it. Through all that lo-fi glitchery, you could see the promise of something marvelous.

That something marvelous came to fruition in *Virtually Dating,* a show that Staake and his company created with Condé Nast Entertainment that streams on Facebook's video platform, Watch. Could there be a more 2018 sentence than *that*? (Just for the record, and what they call "full disclosure": Condé Nast owns *WIRED* and is thus my employer; however, I met Staake long before I knew Condé Nast was involved with the "VR dating" project we first spoke about.) The genius of the show, a send-up of blind-date shows like *Love Connection,* is its kaleidoscopic treatment of reality: strangers meet for a date, but they do it in VR, before ever having met face-to-face. Even better, they're sharing the same physical space as well. Just like in The VOID, they can reach out and touch each other.

This being the early days of this sort of experiment, plenty goes wrong. There's the aforementioned arm through the neck, but the participants are constantly walking through virtual objects, their limbs flailing about in defiance of the motion-tracking modules they have clipped to their joints and feet. The show mines all of it for comedy, transporting each couple from one location to another (the beach! a zombie movie! ancient Egypt!) and shape-shifting each person so many times that Optimus Prime couldn't keep up.

Each episode is only seven or eight minutes long, but they're easily as entertaining as any half-hour dating show, mostly because the participants are so damn *giddy* about everything. VR's current shortcomings make every meet-cute a meet-cuter, it seems. Even better, tactile presence becomes a literal conductor of intimacy. With the glitchiness kicking self-consciousness out the window, touching each other is simply a natural response. It's innocent play, a moment without the *weight* of seduction—yet it still contains all the necessary ingredients.

"The biggest thing to me was simply seeing the power of shared presence," Ryan Staake says. "It's huge. It's beyond the elements of social lubricant and relationship implications, and more that core sense of not feeling like this lonely single person in this vast digital world. You're truly there with another person." Maybe that's why at the end of each episode, when the headsets come off and each person has to decide whether or not they want to go on a regular IRL date, you find yourself rooting for everyone. Dating is already hard; maybe VR can make it a little bit easier.

But don't take my word for it. One Facebook user summed it up more concisely (and profanely) than I ever could. "This is a mile stone [*sic*]," he wrote in a comment on an early episode of *Virtually Dating*. "A staple point in history. when VR is perfected in 20 years, we will look back on this like we do when we watch old school episodes of blind date and laugh whilst playing uber diamond platinum ultra tinder and insta fuck each other."

He's wrong about one thing: it might not take twenty years. Everything else, though? That's up for grabs. Speaking of which, it's time we headed to the culmination of all this intimacy.

# 9

# XXX-CHANGE PROGRAM
## TURNING PORN BACK
## INTO PEOPLE

THE ADULT INDUSTRY may have changed a lot in the digital age, but some things still live up to the stereotype. Like, say, the house in the LA suburbs I'm currently standing in. The freeway looms less than a block from here, shuttling cars throughout the big-box sprawl of the San Fernando Valley, yet we're tucked alongside numerous million-dollar homes, behind a pair of neighborhood-announcing pillars. The Hollywood Bowl is a fifteen-minute Uber ride away. So are the Kardashians.

Inside, things are a little less TV-friendly. Mostly because of the statuesque naked woman on the four-poster bed, but also because of what she's doing. Which involves—and really, there's no more polite way to say this, so bear with me—inserting a string of oversize beads into herself, while a naked guy stands next to the bed, a towel hanging from his erect penis.

195

But August Ames and Tommy Gunn, as fans might know them, aren't here simply to have sex on camera. If that were the case, then the VR camera rig in front of Tommy's face wouldn't be forcing him to lean back so far at the waist that he retires to a daybed in the corner to stretch between takes. ("My hip flexors are killing me," he growls during one lull.) If August and Tommy were making "conventional" porn, then she might not be getting so close to the camera lenses, cooing into them as though they were a lover's eyes. If they were just making your standard wham-bam-thank-you-surgically-enhanced-ma'am porn, with three *X*s but only two lousy dimensions, then the CEO of the company bankrolling this shoot might not be sitting downstairs, having flown here from Barcelona just to be around. And there *definitely* wouldn't be a "clinical sexologist" overseeing the shoot, making sure that the action unfolds in accordance with maximum therapeutic value.

But this is VR porn—in which intimacy is the watchword, eye contact is everything, and studios are seeing moneymaking potential the likes of which hasn't been around since the internet came along and almost cratered the whole damn industry.

Just another day in sunny Encino. Please don't step on the ben-wa balls.

## THE MAN WHO WOULD BE (BADOINK) KING

Todd Glider never meant to get into porn. Back in the mid-1990s, he was the living embodiment of the mid-1990s: a twenty-something with an MFA ("pipe, tweed jacket, all that," he says), living in San Francisco, making zines. Then his girlfriend got a job in Los Angeles, so he started looking for employment in Southern California. One of the listings he saw asked for an "HTML programmer." He got the interview, and the job, but "HTML programming" turned out to be "writing erotic copy for an online adult company."

The job, Glider found, suited him fine—as did the adult industry as a whole. He became the creative director at that first company, then moved overseas to work in Europe. In 2010, he became CEO of a large "digital entertainment company," which now serves as an umbrella over several smaller adult brands. One of those brands is BaDoinkVR, the studio creating the scene that's shooting upstairs today.

BaDoink—and by all means, take a moment to enjoy that glorious name—released its first VR porn scene in the summer of 2015; the company was profitable within a year. It's gone from ten employees to more than ninety, a workforce that is "overwhelmingly coders," Glider says, sitting in the living room of the Encino house. He's sturdily built, with a shaved head and a gregarious mien, and is dressed like he's heading onstage to talk to a crowd of tech developers: dark gray button-down, black pants, Apple Watch. That's not unintentional. The way Glider sees it, VR has the potential not just to make porn profitable again, but to make the tech world respect the adult industry. "This is the first time I feel like we're leading in any way," he says. "Silicon Valley left us in the dust, but now adult is carrying the torch."

Historically, the desire to see naked people doing naked-people things has driven the widespread adoption of otherwise niche consumer technology. VCRs, CD-ROMs, and even streaming video owe much of their early uptake to the fact that they made watching porn more convenient and more private.

But as technology giveth, technology taketh away. The same

streaming video compression that turned YouTube into a juggernaut also robbed the adult industry of a huge chunk of revenue.
Consumers who once bought or rented DVDs could now just go
to so-called tube sites where they could watch high-def porn—
usually pirated—to their hearts' (or other body parts') content. And
for years, studios did what they could to fight the tide, jumping on
whatever technology might help them make some money again:
3-D TVs, ultra-high-def resolution. Nothing worked, because nothing made porn seem fundamentally *different*. At the end of the day,
consumers were watching other people have sex. Nothing would
change that.

Nothing, that is, but the transformative power of presence.

## MERRY CHRISTMAS, HERE'S YOUR FOURSOME

Meet Scott (not his real name). Scott is in his midfifties, married to
his college sweetheart. Lives in the Pacific Northwest, works for a
software company. Scott had never paid for online porn in his life.
Wasn't what you'd call a connoisseur. Didn't know any stars, wasn't
familiar with its various genres. (Yes, there are genres. Please don't
act surprised.) He'd watch some if he was on a business trip, or if
his wife were gone for the day and he was bored. But then Scott got
a mobile VR headset for Christmas. He messed around with the
preloaded games and experiences—hung out inside the Cirque du
Soleil, did some space exploration—and started looking around
for things to do. The first stop was YouTube, to search for VR game
demonstrations.

A funny thing about YouTube's recommendation algorithm:

turns out that if you go looking for VR stuff on YouTube, then the site will start suggesting other videos it thinks you'll enjoy. Eventually, you'll likely come across a "reaction video," as it's called, in which people watch VR porn for the first time. It's a split screen: on one side you see an appropriately blurred scene of what the user is seeing in their headset, and on the other you see the headset-wearing viewer, their expression likely somewhere between shocked and amused. *Interesting*, thought Scott. *Maybe I should check that out.* So Scott went looking for *that* and soon found a free full-length sample from a studio called VirtualRealPorn. (It's no BaDoink, but what it lacks in creativity it makes up for in clarity.) Downloaded it to his phone, popped that in his headset. This wasn't meant to be an erotic journey of self-discovery; Scott's clothes stayed on, and he didn't even touch himself.

The video in question, "Your Neighbors and You," is twenty-nine minutes long. That was an eternity when it was released in 2015—especially in nonporn VR, where most videos were so concerned about motion sickness that they lasted fewer than ten minutes. When the scene begins, you find yourself in bed; a male body stretches away from you, its lower half covered by sheets. The scale is a little strange if you're sitting up in real life, the way Scott was, but you still realize that the body is supposed to be yours, despite the fact that you can see only its stomach and southward. But don't think about it for too long, because three women just walked into your room. "He's even cuter in real life than he is from across the road," says one with a British accent. "Shall we wake him up?" The other two women crawl onto the bed and put their faces close to yours. "Good morning," one of them whispers in your left ear. She giggles. The sound is incredibly close to your ear, and incredibly lifelike.

The way things proceed from there is, in one way, utterly predictable: the four of you have sex in every conceivable permutation. Gymnastic carnality aside, though, you're struck by certain aspects of it. For one, the male performer—i.e., you—doesn't move. Like, at *all*. The hands reach out a few times to hold and squeeze various things, but the actor is for all intents and purposes completely passive. Stranger still, the three women stare into the camera nearly the entire time (at least when they're, uh, facing the camera). If this were conventional flat video, the constant fourth-wall-breaking might seem strange. In VR, though, the effect turns from a gimmick into . . . well, into a moment-maker. Depending where you look, and how skilled an actor the performer is, you really do feel like you're staring into each other's eyes.

Scott watched all this in amazement. He had two thoughts. The first was *This is an experience I was* not *expecting.*

The second one was *I want to see more of it.*

Scott wasn't alone, in either his curiosity or his timing. On Christmas Day 2016—when he got his headset—online tube site Pornhub saw its VR video views jump from around four hundred thousand a day to more than nine hundred thousand. But Scott wasn't looking for the short preview snippets available on tube sites; he wanted the real thing. "I just couldn't believe the immersion level that it provided," he says. "Even though it was a little fuzzy, everything made me realize that this is more than just watching a video in 3-D. When a woman comes up close to your face, you can feel the heat coming off of her, you imagine that you feel her breath. Your brain is tricked into senses that aren't there because of the ones that *are* there."

So he researched some more, tried out the offerings of a few VR studios. Eventually, he settled on a site called WankzVR. (Ba-

Doink, you have a challenger!) He liked that its videos had a sense of humor; there was a *Game of Thrones* parody and a zombie-themed scene that had come out around Halloween. Most of all, though, he says, he liked that "the biggest emphasis seems to be on making it real. Making it intimate."

Adult studios, and the consumers who congregate on their message boards and on Reddit to share feedback, call this "the girlfriend experience," or just "GFE." And for most companies that jumped into making VR porn, it became the watchword. "That was on the top of everyone's list when we first started making content," says Anna Lee, the president of adult VR studio HoloGirls. "'Give me a girlfriend, make me believe that she wants me, make her look at me, make her be intimate with me.'"

But perhaps no one made intimacy the cornerstone of their productions the way Wankz did. The studio created a camera rig that let performers come *riiiiiiiight* up to the lens so they could pretend to kiss it. More and more of the company's scenes became dedicated not to the sex act itself, but to foreplay and face-to-face interaction: whispering, teasing, eye contact. It even started filming the actual sex in a way that explicitly cropped out penetration. Male viewers might, for example, see themselves having sex with a woman in the missionary position—but the camera is tilted such that if they look down, their view stops just below the woman's waist. (Some VR video, including the majority of adult VR, is filmed in 180 degrees, which leaves some areas of the virtual sphere blacked or grayed out.) The focal point, instead, is on the woman's pleasure: her facial expressions, the sounds she makes, the way she moves.

It's hard to overstate how fundamental that shift is. For tens of thousands of years, the vast majority of erotic art has depicted sex in a single fundamental way. Regardless of the surroundings, regardless of the position, regardless of what body part is contacting what body part, sex has defaulted to . . . well, at least one person's genitals being stimulated. With what Wankz and other VR studios are doing, sex in VR becomes not action, but *reaction*. The action, studios know, is already taken care of. It's happening outside of the virtual bedroom, with the viewer taking matters into his or her own hands.

## MARS, VENUS, ETC.

Let's take a moment here to address the obvious: this is all sounding pretty one-sided. You can find VR porn that puts you in a woman's body, and there's stuff targeting gay men, but much like the adult industry as a whole, the vast majority of adult VR content has catered to heterosexual males. To a large degree, that's simply where the money is: according to Pornhub's data, male visitors are 160 percent more likely to watch VR content than women. In fact, women constituted a mere 26 percent of Pornhub's overall visitors in 2016.

Consider, though, the idea that VR porn may be more welcoming to women by default than conventional porn is now. After all, technology may have had an effect not just on the adult industry's financial well-being, but on its sociological one as well. After high-speed internet and streaming technology gave way to tube sites, the rise of smartphones made porn more snackable than it had ever been. In 2015, mobile users accounted for 53 percent of Pornhub's traffic, and that majority has only grown. (In the States, the curve is even steeper: in 2016, 70 percent of Pornhub's US traffic came from mobile users.) On average, people spend fewer than 10 minutes per visit to Pornhub—enough time to find the clip they want, handle their business, and move on.

Much like the rise of MP3s created a precipitous dip in album sales, so too did plentiful (and free) porn clips replace full-length movies as the conventional unit of consumption. Whether by cause or by correlation, the porn industry was becoming a buyer's market, and the economic impact trickled down to its workforce: performers earned less and less for sex acts. Those sex acts, in turn, were filmed in increasingly demeaning ways. A 2010 study of three hundred popular porn videos found that 88 percent of them featured some degree of physical violence toward women such as slapping, spanking, or gagging. That's not to kink-shame or to deny that a spank can't be pleasurable for both parties, but when everything was available for free, extremity became the way to stand out.

Virtual reality, though, has the potential to reverse that trend through the magic of empathy. With the frame gone, the viewer is *in* the scene. And once you're in the scene, thanks to presence, you're no longer a voyeur. You're a participant. No more detachment, no more desensitization.

Does that make things more arousing? More difficult? Awkward? Embarrassing? It depends on the scene. It depends on the person watching it. But regardless, that potential to implicate the viewer, to put them on equal footing with the fantasy they wanted to see, promises to upend pornography in a way no one has considered.

We're already seeing that in the early years of VR porn. "People are responding to what's kind of the antithesis of traditional porn," says Doug McCort. And if anyone should know, it's him.

For the past two years, the forty-six-year-old Alaskan has been reviewing almost every VR porn scene that's been released online for his website 3DPornReviews.com. And I don't mean he watches a scene and then does his best Roger Ebert impression; I mean he *really* reviews it. He watches it once, to make sure it's worth writing about—there are so many releases these days that he has to be a bit more discerning—then watches it again, pausing and unpausing, grabbing screenshots for his readers, taking his headset on and off. All told, each review takes four to six hours on average, sometimes more.

"Porn had kind of degenerated into gonzo-type shit," McCort says, using the industry jargon for hardcore, pretense-free porn. "Where else are you going to push it after forty or fifty years? All you can do is push through visual extremes or physical extremes, and that's silly. VR offers access to things that you just can't get in porn. You're seeing a throwback to the basic things that are erotic when you're in close proximity to another member of the opposite sex. It's far from pornographic and much more like human intimacy."

Performers have become more than fantasy objects; they've become fantasy partners—and more important, they've become *people.* "I find that I care more about the people in the scenes than I used to," Scott says. "Even though they're still playing themselves being a porn star, their personality comes through in a way that I find kind of fascinating—so I actually seek out behind-the-scenes interviews or podcasts where they're guests, just to hear a little more about their life."

Sounds like a crush, doesn't it? Scott maintains, though, that VR porn has actually rekindled his connection with his wife of nearly forty years. "My interest in sex with my wife has increased significantly," he says. "She thinks it's because I got a different job and I'm less stressed out, but it's actually because I realized how enjoyable my intimacy with her is. When I first started watching VR porn, I thought, 'Maybe this is an opportunity to fantasize about one of the women I had this experience with in VR when I'm with my wife.' That did *not* work. There was this cognitive dissonance that actually made it worse. Focusing on my wife as my wife, the person that loves me and that I love, was so much more satisfying and exciting—even though I had this separate set of experiences in VR that maybe made me interested to have sex that night." VR might have been an aphrodisiac, in other words, but it wasn't an alternative.

## VIRTUAL SEX,
## BUT REAL INTIMACY

Back at the BaDoink shoot, not everyone is a VR vet. This is the first time for Tommy Gunn, the male performer—but since he's appeared in more than seventeen hundred films, it's going to take more than some fancy cameras to faze him. "From what I understand," he says, "I just have to lay back and enjoy the ride. It's better than a sharp stick in the eye."

Gunn looks a little like Bronn, the roguish sellsword from HBO's *Game of Thrones*—if Bronn had grown up in New Jersey and liked customizing military vehicles. He shares Bronn's plainspoken manner as well. "Porn is, at the end of the day, a penis in a hole," I hear him tell a crew member in the kitchen before the shoot. "That's what it is."

Not necessarily in this case, though. For one, there are those aforementioned ben-wa balls, which at least for the moment are occupying August Ames's attention. ("Looks like a cat toy," Gunn says from the daybed between takes. He's part right; however, being bright green with red, they also look like two tiny heirloom watermelons.) And they're just one of a number of things different about this movie-to-be, which is BaDoink's attempt at using VR to make adult content that's both intimate and instructive.

Like many other VR porn pieces, "Virtual Sexology" is being shot from the first-person perspective of the male performer (in this case, Gunn). This perspective isn't new to porn; POV (point of view) is now a genre all its own. But it's close to the default treatment in VR, at least in the technology's early days, because it creates so much face-to-face intimacy for the viewer. Also, as in many other VR porn experiences, the director tells Gunn that he needs to remain mute and largely still. That can lead to some odd contortions, since the camera rig needs to be placed at his eye level without interfering with what I'm just going to call his operational appendages. As uncomfortable as that may be for him, though, it's a must for VR—because it forges a stunning link between the viewer's brain and the body that presence tells them they're occupying.

But more than the camera angle, it's the very structure of "Virtual Sexology" that makes it unlike just about any VR porn movie (or porn movie, period) out there. That starts with the sexual encounter itself. Although it includes most of the menu items you'd expect, it's more like an instructional video: throughout, August Ames looks into the camera and coaches "you" (as embodied by Gunn) through various techniques that range from deep breathing to ways to delay orgasm.

## Skeptic's Corner: Embodiment

**You:** So you're telling me that if I look down and see a naked body, I'll just think it's mine?

**Me:** Not at first. And sure, rationally, you'll obviously know that you don't have Tommy Gunn's abs. Or his, well, professional equipment. (Seriously, you could pick it out of a lineup.) But if you're ever in a mood to go reading through VR research . . .

**You:** I'm not. That's why you're here.

**Me:** Right! That's why I'm here. Okay, so what we're talking about dates back to a concept known familiarly as the "rubber hand illusion." In 1998, a psychiatrist and psychologist in Pittsburgh asked volunteers to sit with their left arm on a small table and then placed a screen between each subject's arm and body so that they couldn't see their own arm. They placed a life-size rubber arm on the table and told the volunteers to focus on the arm—then stroked both the hidden real hand and the rubber fake hand with paintbrushes, timing the strokes as closely as possible but locating the strokes in slightly different places. They did this with ten volunteers. Eight of those ten wound up feeling as though the rubber hand were their *actual* hand: they felt the paintbrush in the place where they saw the rubber hand touched or thought that the stroking they felt was caused by the visible paintbrush.

Ten years after that experiment, a different group of researchers in Spain replicated the rubber hand illusion in VR, calling it the "virtual arm illusion." In their own words, they were able to induce "a feeling of ownership of simulated body parts in a virtual environment."

And now, the virtual arm illusion is one of the foundations of intimacy in VR porn. Remember Scott, the guy from earlier in the chapter? Here's how he put it to me: "If I reach out with my hand and I superimpose my hand in the same position that the actor's hand is, it triggers a response in my brain that that's my hand. Being in the same space really, really transforms the experience to the next level of me being there."

**You:** Sure, but what if I'm a woman and I look down and see a man's body? Or I'm a man and I see a woman's body? Presence has its limits, right?

**Me:** You'd think so, but it's not that easy. Not long after those researchers in Spain proved the virtual arm illusion, another team led by the same person—a scholar named Mel Slater, whose presence research has defined much of the field—wanted to see whether they could induce the feeling of a full-body transfer. They created a virtual environment in which male volunteers sat and looked across the room at a woman stroking a young girl's shoulder. After two minutes, the scene changed and the men's perspective shifted. Some of the men were given the first-person perspective of the young girl; if they looked down, they saw the young girl's blouse and skirt, and if they looked in a mirror, they saw the face of the young girl. The other men found themselves between the girl and the woman, but not actually inhabiting the body of either.

For the next few minutes, the men watched the woman stroke the girl's shoulder, while a researcher in the lab stroked their own real-life arm. Suddenly, the men's perspectives changed again, so that they were floating above the woman and girl; they then saw the woman slap the girl, three times, across the face—slap! slap! slap! They then returned to their previous position, and the experiment ended.

Afterward, the men filled out questionnaires, but the researchers were doubly interested in what happened to the men's heart rates. In particular, they looked at how men's heart rates slowed when they saw the woman slap the young girl; heart rate deceleration has been linked with so-called aversive stress, or the desire to escape a situation. The men who had been given the first-person perspective of the young girl had a significantly stronger physiological reaction both to seeing the young girl get slapped and to returning to the young girl's body—just as if the woman posed a real threat to them.

"Through an IVR [immersive virtual reality] a person can see through the eyes and hear through the ears of a virtual body that can be seen to substitute for their own body," the researchers wrote, "and our data show that people have some subjective and physiological responses as if it were their own body." And not just a different body—a body of a different gender, and even a vastly different age.

All of which is to say: you might not have a porn star's body, but VR can make you think you do. ∎

The actual script is mostly voiceover—in the finished film, a female narrator will handle the exposition, providing somewhat clinical commentary on the benefits of Masters and Johnson's "squeeze" technique, or how various positions can maximize stimulation for either or both parties—which makes August Ames's dialogue sound like a cross between an overacting porn star and an everyone-gets-a-trophy preschool teacher. "Oh my God," she coos breathily into the camera during a segment on Kegel exercises, as Tommy Gunn does what can only be described as dong push-ups, a towel draped over his laboring penis. "Your dick is *so strong*."

If you think that sounds cheesy in writing, imagine standing on a powered-down treadmill in the fully furnished rental house, a scant ten feet from the bed, scribbling notes furiously. And if you think it sounds like another sad example of the adult industry peddling the myth of a subservient yet sexually insatiable woman, you're . . . well, you're not wrong. At least not in this case. But the porn industry flocking to VR isn't confined to heteronormative, male-first fantasies. After BaDoink released "Virtual Sexology," in fact, it went on to make a sequel shot from the female performer's point of view. Another site, VR Bangers, recently released a male-female scene that was shot from both the man's perspective and the woman's, and then synchronized—with the hope that a couple at home will pop on their headsets, reenact what they're seeing, and enjoy being other people for seventeen minutes and twenty-six seconds. (Or however long it takes before they, uh, get bored.) There's gay VR porn, trans VR porn, BDSM VR porn; basically, find a flavor, stick a "VR" in the middle of it, and it likely exists. And if it doesn't, it will soon.

Back in the very beginning of the book, I took some time to set up the idea that real life didn't always care about reporting. Startups can fail; businesses can change names, be bought or sold, or simply go under; people can change jobs, even careers. That sort of unpredictability has a tragic side as well. In December 2017, during the book's final editing phase, we learned the horrible news that August Ames had taken her own life. She was twenty-three years old. ∎

In 2011, two computational neuroscientists analyzed more than four hundred million online porn search results and unpacked what they found in a book called *A Billion Wicked Thoughts: What the World's Largest Experiment Reveals About Human Desire*. In an interview with *TIME*, coauthor Ogi Ogas explained the gender gap thus:

> Women prefer stories to visual porn by a long shot. . . . There are two reasons. Both come down to fundamental differences between the male sexual brain and the female sexual brain. One of the most basic differences is that the male brain responds to any single sexual stimulus. A nice chest, two girls kissing, older women—if that's what they're attracted to. Any one thing will trigger arousal in a male. Female desire requires multiple stimuli simultaneously or in quick succession. It takes more stimuli and more variety of these stimuli to trigger genuine arousal. For a guy, the most common form of [masturbation material] is a 60-second porn clip. For a woman, it can be a 250-page novel or a 2,000-word story. That's the way to get multiple stimuli. Stories have greater flexibility to offer a greater variety of stimuli. In male erotica, sex appears in the first one-quarter of the story [or film]. For women, it's halfway in. There's more time to develop the character before sex.

But if anything can close that gap, it's virtual reality. VR porn already shares more hallmarks with the longer stories Ogas describes than with a sixty-second clip. "It's really hard to get somebody to sit down and actually consume an entire piece of adult pornography," says Doug McCort, the omnivorous porn reviewer. "But for some reason it's working in VR. People are digging the entire experience." Some of that is practical, he allows: "It feels weird to go through the hassle of strapping up and putting the [headset] on just to knock out a ten-minute clip."

But even if VR porn scenes last the same amount of time as a conventional scene might (and many do, when you're comparing them by original length rather than the clips that filter out from pay sites to tube sites), they *feel* different. That's because of the eye contact, and the whispering, and the simulated kissing, and all the ways that directors and performers are learning to take advantage of proxemics and presence. They also feel different because the entirety of the scene, from premise to climax, feels like part of the same encounter.

As the medium in which we watch sex has changed, the sex itself has become increasingly decontextualized. When VCRs moved porn consumption from theaters to homes, movies went from long features to collections of scenes; when the internet came around, those scenes shortened to clips. And when smartphones emerged, those clips shortened even more: porn GIFs have become astoundingly popular, especially among young women. Yet, VR reinstates the totality of an intimate encounter, even if it's aimed at men who might otherwise treat porn as instant gratification. There's a seduction to these scenes when consumed in their entirety; the moments *before* sex matter again.

Keep in mind, this is all for prefilmed experiences, which means that this is virtual sex defined by its limitations. It may or may not be fully compatible with a given viewer's tastes. It conscripts the viewer, but only into a body that isn't actually theirs. This is virtual sex in which viewers are rendered dumb and mute: they can speak, but who's going to hear them? They can only be a consumer, rather than a true participant. (There are, however, app- and internet-connected vibrators and masturbation devices that can be synced to what's happening on the screen. These so-called smart toys get one step closer to tactile presence in VR porn.) Make no mistake, presence can enable some miraculous reactions, but a one-way transmission like this stops short of full-fledged reciprocal intimacy.

That's already changing, though.

## SMILE, YOU'RE ON CAMERA

"I'm going to pee, because I'm a pee monster. That's my official title." Ela Darling gets off the couch for what seems like the third time in twenty minutes and heads to the bathroom. We're sitting in a hotel room in Miami's South Beach neighborhood. Outside, it's hot. Downstairs, the hotel's outdoor pool and bar throng with women in bikinis and jacked dudes in jean shorts and tribal tattoos. Some of them might be here anyway—this is Miami, after all—but most of them are here for XBIZ.

The AVN Adult Entertainment Expo is a porn trade show people have heard of. XBIZ is . . . not. XBIZ is a business conference, filled with webmasters and affiliate marketers. It's the nerds' table in the middle-school cafeteria of the NSFW world. And Ela Darling might be a porn performer, but she's also a die-hard nerd, and these are her people.

As a kid, Darling fell in love with the idea of virtual reality. This was the late 1990s, early 2000s; *Johnny Mnemonic* and the Nintendo Virtual Boy had already come and gone, and VR had gone from brain-busting sci-fi concept to schlocky punch line to faded cultural footnote. But still, Darling was an avid reader and D&D player, and the idea of getting lost in an immersive world—"making visual what I was already losing myself in books for," as she puts it—was something she found not just exciting, but romantic.

Not surprisingly for an active reader, Darling got a master's degree and became a librarian. Perhaps more surprisingly, she then stopped being a librarian and started acting in pornographic movies. (Yes, that means she officially became a sexy librarian. Fun fact: she has the Dewey decimal number for the Harry Potter books tattooed on her back.) And after a few years of bondage scenes, masturbation videos, and girl-on-girl movies, Darling attended the E3 video game trade show and tried an early version of the Oculus Rift. "The first thing I think of when I hear of new technology," she says, "is 'How can I fuck with it?' or 'How can I let people watch me fucking on it?' Usually there's one or the other application if you think hard enough." With the Rift, Darling didn't have to think too hard at all; now, she's at the forefront of the world of VR camming.

If regular porn is a movie, camming is closed-circuit TV: performers stream live shows to their webcam, which people watch via their browsers on cam sites like Chaturbate or LiveJasmin. It's been around for years, though with the proliferation of good cheap cameras and high-speed internet, it's become an increasingly popular part of the adult industry.

It's also become a lucrative proposition. Because it's a live show, it's less prone to piracy than conventional scenes. Additionally, most cam sites generate revenue not from advertising or subscriptions, but from tips. Users can watch shows for free but are able to send small micropayments to performers—"for a good show, sending fun gifts and requesting sexy one-on-one Private Shows," as website Cam4 describes it—in the form of tokens they buy from the site. While the cam sites keep a portion of those tips, the business model is more like a small brick-and-mortar business than the movie business: performers pay overhead to the cam site in order to control their own destinies.

Camming wasn't Ela Darling's first VR play. In fact, VR camming didn't exist at all in 2014. That's when she read a Reddit post
from two guys who wanted to make VR porn. *She* wanted to make
VR porn, so they flew her from California to Maryland. In true tech
start-up fashion, they turned out to be twenty-one-year-old college students. ("It was very *Weird Science*," Darling says.) Nonethe-
less, they shot a test scene in their dorm room. Rather than invest
in an array of pricey high-end cameras like other fledgling VR
video companies, they went decidedly DIY, taping together two
GoPro cameras to create a 3-D image on the cheap. (Again in true
tech start-up fashion, Darling wore an R2-D2 swimsuit [at least
initially].) After she flew back to LA, one of the students emailed
her; he'd finished processing the test scene and was so blown away
by the footage that he wanted her to be a partner in the venture.
"This is unlike any porn I've seen," he wrote to her then. "It's like
I'm watching an actual person." (That's a lot of quote to unpack.
The subtext suggests that things critics say about porn are true—
that it literally dehumanizes its actors. That's not a topic for this
book, but suffice it to say that whether or not we're talking about
conventional porn, VR's unique qualities cast that sentiment in a
different light altogether.)

Now, the two are roommates and business partners in LA.
They've gone from GoPro cameras to a sophisticated custom-
built rig that lets performers get incredibly close to the lens—*the
better to seduce you with, my dear!*—and Darling has gone from
filmed content to camming. Well, partially: this morning, before
we met up, she filmed a scene with another female performer
she's friends with. ("She fucked my butt!" Darling says, in a non-
chalant singsong that sounds like she's telling me the copier is

broken again.) She's also gone from doing weekly live cam shows on her own website to being the head of VR for cam site Cam4. Cam4 is the ninth biggest adult site in the world, and the new platform has boosted her visibility significantly. When she did her first show for the site, hundreds of users were watching at any given time.

But despite those numbers, the experience maintains its illusion of one-to-one communication. Darling is careful to use singular pronouns only when she talks to the camera, and even addresses individual users by name when they type something into a chat window. "Cams are about establishing an intimate relationship and a shared sense of vulnerability," Darling says. "When I'm camming in VR, I actually feel a lot more engaged, because I know that for people watching me in virtual reality, their experience is dominated by me. They're not checking their email, they're not texting, they're not fucking making lunch. They're only looking at me, and that is really powerful. Like, knowing that I'm conquering all of your attention right now? Damn."

That direct connection can engender some surprising effects. "There's a sense of reciprocal attraction which you don't get from porn," Darling says. "I'm being naked and masturbating on camera and you're watching it, and that makes you feel you can kind of open up a little bit. When I cam in virtual reality, guys get to a point of personal self-disclosure much sooner than they do on a regular cam platform."

Like a lot of cam performers, Darling has regulars, viewers who come back anytime she does a show. One week, she noticed that one of her recent regulars was missing. He was back the following week, and he asked her for a private show. He and his girlfriend had broken up, he told her, and he just wanted to be with someone he felt he could trust. "He'd only seen me a few times," she says, "and he felt that VR was a safe, comfortable place to retreat to when he was feeling vulnerable."

"There's a relationship there," she continues, "because they feel like they're in my bedroom. And to some degree they are." While some cam performers stream from studios built for that specific purpose—a house full of subdivided workstations, each furnished with a bed and a webcam—Darling just sets up a VR camera rig in her own room. When you're there with her, you're really there, from the skulls strung up above her bed to the bejeweled gas mask on a female mannequin torso to the many skulls on her bookshelf. (Homegirl is into skulls.)

That relationship, such as it is, *matters*. And as much as it matters to conventional cam customers, who often pay extra for private shows with performers and buy them things from their Amazon wish lists, it matters even more in VR. "A lot of people don't have access to romantic relationships," Darling says. "Maybe they have mobility issues, or health issues, or they work a lot, or they're just really socially awkward, or any number of reasons why someone might not be able to have that kind of relationship. This is giving them an opportunity to interact with people they wouldn't otherwise be able to. That's huge to me."

But what if that person is already *in* a relationship? Darling's regular had broken up with his girlfriend; had his girlfriend known that he spent time, and possibly money, watching another woman masturbating in real time?

The question of whether watching porn constitutes infidelity simply doesn't have a single answer: every couple is different. Yet, the added power of presence introduces a new complexity to the question. Is getting aroused by the depiction of another person different when that depiction *expressly creates the illusion that you're really there with that person?*

Remember Scott, the fifty-something software guy? A few months after the last time I talked with him, he sent me a long, thoughtful email. During his time on the message boards of his favorite VR porn site, he wrote, he had read other men's stories about watching VR porn with their wives. Some had simply shown their wives movies shot from a woman's perspective, and others had had sex while wearing a headset. Scott began to wonder whether such a thing was possible in his relationship. VR porn had already rejuvenated his sexual connection with his wife; why couldn't he tell her about it? So he did.

It didn't go well. She asked to see one of the scenes he had watched—and after watching the entire scene, she said, "This feels like adultery." Scott was shocked. In his mind, VR porn was simply fantasy, albeit particularly vivid fantasy. His wife, though, reminded him of a book they had read together that stressed the power of visualization: if you practice shooting free throws in your mind, for example, your subconscious will eventually internalize the repetition, leading to real-world improvement. VR porn, his wife said, was similarly conditioning his mind to have an affair—it was an adultery simulator. (The fact that Scott had researched his favorite performers didn't help either; they were becoming like girlfriends to him, his wife said.)

Scott didn't think that he would ever cheat on his wife, but he canceled his WankzVR membership—and began to think critically about his own VR consumption:

> I had convinced myself that VR porn was different—it was
> good porn. Regular porn depicts a man (or men) carrying

out acts on a woman, sometimes in very degrading ways. In VR porn, the woman is generally seducing *you,* making love to *you.* She's the one in control, she's empowered. The scenes tell a story and there is some thought that goes into the dialogue and plot. The sex itself tends to be more tender, more real, with longer foreplay and more realistic endings. I reasoned that if VR is rewiring my brain, at least it's very close to the real experience of sex.

In hindsight, I now realize that while VR porn *is* different in many ways from traditional "2-D" porn, the high that it can produce is significantly more potent and thus more dangerous. The combination of a realistic 3-D environment containing a person who is focused on pleasuring you triggers dopamine spikes that flat porn can't touch. I remember the first time a girl whispered in my ear in VR—I could swear that I could feel her breath and the heat of her cheek radiating against mine. It sent tingles down my spine. That feeling lessened in later videos, so I realize that my brain was, in fact, getting used to the experience (desensitization).

"I feel like I've dodged a bullet by coming clean to my wife when I did," he wrote; he was excited to rediscover "authentic sexuality" with her.

I'm happy for Scott. Happy that he was honest with his wife; happy that she was honest with him; happiest of all that they found a way to move forward. "So many people don't have a conversation about what the boundaries of their relationship are," Ela Darling says, "that when someone ends up doing something that makes someone else feel violated, both people are upset: 'You just did something that made me feel like you violated my trust.' 'You didn't communicate to me that this was a boundary for you, and now you're mad at me for something I had no idea would hurt you.' If VR can encourage people to have a fucking conversation with each other, that's awesome."

But the intersection of VR and eroticism is just beginning. We'll continue to see the adult industry create content to match the capabilities of the most powerful headsets, just like it's always been at the forefront of technological adoption. As VR reaches everyone, we'll see studios like BaDoink and Wankz and Yanks and Kink.com and Naughty America and a dozen others broaden their offerings, catering to the panoply of tastes. We'll see camming improve, with cameras that allow viewers to lean in closer to the performers they see in their headsets, and we'll ultimately see performers wearing their *own* headsets, connecting with paying customers for private time (as avatars only, of course). And just as certainly, we'll see handwringing about how the immersive qualities of VR porn make it a danger—to young people, to women, to relationships, to the fabric of society itself. 'Twas ever thus, right?

But consider this: the intimacy Scott and his wife share now is stronger than it's been in years. And that's thanks to VR.

# 10

# WHERE WE'RE GOING, WE DON'T NEED HEADSETS
## LET'S GET SPECULATIVE

A FEW MONTHS AGO, I was sitting at work when I had a sudden urge to know how far apart California and Finland were. Before you bemoan the American public educational system, I promise I know where Finland is. It's just that an acquaintance had been telling me about a burgeoning VR development scene there, and I've always been curious about visiting Scandinavia, so I'd been kicking around the idea of a trip. I hate long flights, though, and so what I really wanted to know is the most direct path a plane could take to get to the land of reindeer and *kaalikääryleet*. (Full disclosure: I don't know what that is, I just like the way it looks like it might sound. Two umlauts in a row!) The internet can do a lot of things, but it can't beat a globe, so I reached out a hand and pulled one of those off the shelf that was floating to my left.

When I dragged it in front of my face and let go, the effect was instant: the Earth was floating directly in front of me. So far, so good; the daylight that was coming through the office windows wasn't so bright that it affected the clarity of what I was looking at. The planet was a little small, though, and I wanted to make sure I could see Finland clearly. So I grabbed the globe with my two hands and pulled them apart to make the globe expand—then pulled my right hand toward me and pushed my left hand away to spin the globe westward. I leaned forward a bit to get a better look at the charming little peninsula, but noticed some movement in the corner of my vision. It was my co-worker, trying to get my attention.

"What's up?" I asked.

"Sorry to interrupt," he said. "I just never know what you're doing in there."

"Just reading an email!" I said. I held out my hand again, grabbed the Earth, and moved it out of the way so I could see my co-worker better. He wouldn't notice either way; it just felt a little bit rude to have a world hovering between us.

Assuming you've been paying attention for the past two-hundred-odd pages, something about that story likely seems off. Maybe it's the way I was grabbing objects with my bare hands, or maybe it was the fact that I could see the real world—daylight, my co-worker—but it just doesn't seem like a typical VR experience.

That's because it's not. This isn't VR at all; it's the other side of the reality coin, *augmented* reality, courtesy of a headset called the Meta 2. The headset doesn't use a screen, but rather a see-through visor that lets me harness virtual objects like browser windows and 3-D globes without becoming oblivious to the outside world. When I put it on, its cameras scan my desk and create a map of my space; if I move something virtual behind my Iron Man pencil mug, then the mug actually blocks it.

Augmented reality, or AR, is still a few years behind VR, despite generating an insane amount of momentum in the past couple of years. But when you put the two together, we're talking about the ultimate conclusion to all of this: a single device that can mix the virtual and the real in varying degrees to create an experience that can range from life itself to pure fantasy. And if you thought VR alone was cool, you ain't seen nothin' yet.

## AUGMENTING YOUR WORLD

In its simplest form, augmented reality simply means placing some sort of visual information on top of your usual view. VR shuts out the world, replacing it with an all-encompassing artificial world; AR starts with the real world and then adds new stuff to the mix. Remember Ivan Sutherland's Sword of Damocles from all the way back in Chapter 1? You could see through that. In other words, it may have been the first head-mounted immersive computing device, but it was really more augmented than it was virtual.

The concept of enhancing your world with visual overlays has been around for decades. Some people even credit *The Wonderful Wizard of Oz* author L. Frank Baum with the idea more than a century ago; his 1901 novel *The Master Key* features "the Character Marker," a pair of glasses that enables its wearer to see letters on other peoples' foreheads. The phrase "augmented reality," though, is only about as old as *Snow Crash,* having first been coined by two Boeing researchers in the early 1990s.

Regardless of how old it is, AR has been part of your life for longer than you likely know. When the 1996 NHL All-Star game made the hockey puck glow blue so your eyes could follow it better? That was AR. Yellow first-down lines in football games? Strike zone graphics in baseball? NASCAR races that show cars' speeds in real time, superimposed above the pack? AR, AR, and AR. (Actually, those are all the work of the same company, Sportsvision, the cofounder of which was involved in developing the glowing puck.)

AR for the masses took a big step forward in 2011, when Nintendo released its 3DS handheld gaming device. The 3DS had two little cameras on it that pointed outward; if you aimed them at special playing cards that you laid on a table, on the device's screen you'd see 3-D creatures crawl out of them and onto the table. Imagine that final scene in *The Ring*—except instead of a crabwalking dead girl skittering out of a well and into your living room, it's a tiny little Mario emerging and shouting "Whoo-hoo!" It was absolutely mind-blowing.

But Nintendo wasn't done. Think back to the summer of 2016, when you and everyone you knew played *Pokémon Go* for at least a couple of days. That game, a collaboration between Nintendo and an experimental Google lab that spun off into its own company, was essentially a scavenger hunt with an AR twist. Leveraging your phone's GPS, it would alert you when creatures were nearby, and you'd need to peer through your phone to see and capture them. Did it matter that they were, at least where I live, invariably Rattatas and Pidgeys? It did not. The simple joy of seeing these collectible little menaces dotting the sidewalks and parks of my city was like the first glimpse of a future that was coming at us fast. (Not the one in which Pokémon evolve and overthrow humanity—the other one. But I, for one, welcome our Zapdos overlords.)

In fact, augmented reality has swept through the smartphone

world far beyond gaming. You might not use or, depending on your age, even *understand* Snapchat, but the app's silly camera filters—the ones that turn you into a dog or let you and your friend swap faces—are a prime example of AR. Last year, Apple and Google both released a set of tools so that developers could create mobile apps; Facebook did the same. The early experiments have been remarkable, bringing *Wizard of Oz* characters into the physical world and letting you fly a virtual remote-controlled plane in the air around you. Yet, the most life-changing are also the most pedestrian: Why waste time looking for a tape measure when you can just aim your phone at a surface and have a virtual ruler appear on the screen, magically extending as far as you need?

As well-suited as your phone may be to delivering simple AR, though, it falls well short of delivering any sort of presence. What you see on its screen might be novel, even magical, but it's quite literally at arm's length. But if you step through that screen—by wrapping it around your eyes, not unlike a VR headset—then all of a sudden that magic becomes something much, much different. In fact, the word "magic" is built into the name of one of AR's largest, riskiest, most closely guarded projects.

## BUT WAIT:
## THE ALPHABET SOUP GETS THICKER

The first rule of Magic Leap is that you don't talk about Magic Leap. The secretive Florida company has raised billions of dollars from titans like Google and China's Alibaba, but if you want to see a demonstration of what the company is working on, you'll need to sign a nondisclosure agreement that threatens you with everything short of fire and brimstone if you divulge any confidential information. I'm one of very few journalists who has used Magic Leap, but before December, when the company finally released images of its soon-to-be-released developer kit, I couldn't even tell you what it looked like. In the interest of remaining uncrisped, I'll simply tell you things that *other* people have seen and talked about. If you happen to infer that I've experienced those things as well, though, that's on you; I can't stop you from jumping to such obviously (wink) erroneous (wink) conclusions (blinkity wink-wink).

Andre Iguodala, a member of the NBA's Golden State Warriors, risked his own NDA to describe having a "character" land on his hand that was so real he could feel its warmth. He talked about using his gaze to control dials and throwing an eighty-inch TV on the wall.

In a cover story for *WIRED* in 2016, Kevin Kelly described interacting with a virtual robot hovering above a desk:

> It is steampunk-cute, minutely detailed. I can walk around it and examine it from any angle. I can squat to look at its ornate underside. Bending closer, I bring my face to within inches of it to inspect its tiny pipes and protruding armatures. I can see polishing swirls where the metallic surface was "milled." When I raise a hand, it approaches and extends a glowing appendage to touch my fingertip.

Kelly saw more as well:

> I saw human-sized robots walk through the actual walls of the room. I could shoot them with power blasts from

a prop gun I really held in my hands. I watched miniature humans wrestle each other on a real tabletop, almost like a Star Wars holographic chess game. These tiny people were obviously not real, despite their photographic realism, but they were really present—in a way that didn't seem to reside in my eyes alone; I almost felt their presence.

Magic Leap calls its technology "mixed reality," claiming that the three-dimensional virtual objects it brings into your world are far more advanced than the flat, static overlays of augmented reality. In reality, there's no longer any distinction between the two; in fact, there are by now so many terms being used in various ways by various companies that it's probably worth a quick clarification.

**Virtual reality (VR):** The illusion of an all-enveloping artificial world, created by wearing an opaque display in front of your eyes. You know this one well by now—if not, might I suggest starting your books at the beginning rather than on page 227?

**Augmented reality (AR):** Bringing artificial objects into the real world—these can be as simple as a "heads-up display," like a speedometer projected onto your car's windshield, or as complex as seeing a virtual creature walk across your real-world living room, casting a realistic shadow on the floor.

**Mixed reality (MR):** Generally speaking, this is synonymous with AR, or at least with the part of AR that brings virtual objects into the real world. (I know this sounds like a cop-out, but it depends on who you ask.) However, some people prefer "mixed" because they think "augmented" implies that reality isn't *enough*.

But just as MR can mean bringing virtual things into the real world, it can also mean the reverse: bringing real things into the virtual world. Back in Chapter 4, we talked about how Lytro used multiple cameras to create volumetric video that you could move within. Mixed reality encompasses that as well: Microsoft uses the term for the way it digitizes humans in order to create "holograms," which can then be used in both VR and AR. (Microsoft calls its VR headsets mixed reality as well, but that seems more about branding consistency than precision.)

And for still others, mixed reality isn't an experiential technology as much as an illustrative one: it combines video footage and VR footage for a hybrid effect that demonstrates what people in VR are actually experiencing.

**Extended or synthetic reality (XR or SR):** All of the above! These are both catch-all terms that encompass the full spectrum of virtual elements in visual settings. You won't hear people using them all that often, but as we'll get to in a moment, it's bound to happen in the coming years.

At this point, I've been fortunate enough to experience four high-end AR/MR headsets. There's Magic Leap, Microsoft's Holo-Lens, the Meta 2 (which I was wearing at work), and a fourth, from the company Avegant, which won't be making its own headset but will be licensing its technology to other companies starting this year. Each of the headsets is mind-blowing for a similar reason: the experience is not like looking at a screen at all. The virtual images are remarkably sharp, no matter how much you enlarge them. That's because the images, in most cases, are projected or reflected directly into your eyes—no pixel-based display is necessary. I've walked around rotating planets, witnessed dogfights between space vehicles, held wild animals in the palm of my hand, and battled robots. In the Meta 2, I could open a web browser window and make it as large or small as I wanted and move it anywhere in my field of view; no matter what I did, the text was perfectly legible.

If VR is a stumbling toddler, though, then AR/MR is a third-trimester fetus: it may be fully formed, but it's not *quite* ready to be out in the world yet. The headsets are large, the equipment is far more expensive than VR (the for-developers-only version of the HoloLens sells for three thousand dollars), and in many cases we don't even know what a consumer product looks like.

Still, we're rapidly approaching the intersection of AR and VR. After steamrolling AR onto its iPhones, Apple reportedly is hard at work on an AR headset to be released in 2020. AR headsets like the Meta 2 and Magic Leap's system can theoretically *become* VR headsets by turning their visors or lenses opaque to block out the real world and projecting a fully artificial environment into users' eyes. On the VR side, Oculus has made no secret of its plans for future headsets to be able to instantaneously see and map your surroundings and then render them inside your headset.

Meanwhile, all the things that make these headsets bigger than we want them to be—optics, displays, processing power, battery technology—continue to improve. There's not a VR or AR company in existence that doesn't use the "pair of glasses" comparison to talk about what these things will look like in the future. And depending on who you talk to, that future might be sooner than you think.

One of those people is Edward Tang. Tang is the cofounder of Avegant, one of the companies working on that mind-bending variety of AR/MR. One sunny June afternoon last year, I visited Avegant's office, and Tang took me through a series of experiences using the company's prototype headset, all of which were stunningly clear and immersive. I walked through a giant room-scale solar system; when I got close to Saturn, it started blasting Beyoncé's "Single Ladies." (You liked the planet? Then you should've put a ring on it.) Thanks to proxemics, I walked up to a virtual woman, close enough to count her eyelashes—an experience I didn't exactly enjoy. Oh, and here's me holding a sea turtle in my hand.

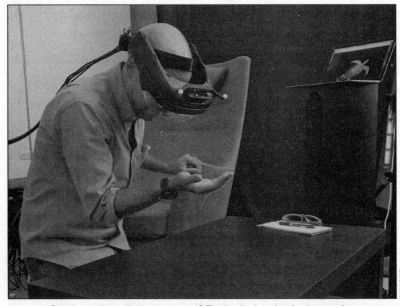

See the turtle on the laptop screen? That's what's swimming in my palm.

Before you say it: I know. I know how big and unwieldy that thing looks, and I know there's a bunch of cables coming off the back like I'm Keanu Reeves waking up in *The Matrix*. (Despite the umbilical-Neo haircut, I promise that those cables aren't coming out of my actual head.) No matter how cool the demo was— and the sea turtle was only one tiny part of it—there's no getting around the fact that this is a work in progress. So afterward, sitting at a table with Tang and the company's CEO, I started to ask about what Tang saw as the end point of the company's work. "So when you think of your dream device . . ."

Without a word, the CEO held up a pair of eyeglasses. "Yeah," said Tang. "That's where we want to try to get to."

I'd heard that only about eight million times over the previous five years, so I pressed him a bit. "What kind of a time frame, though?" I asked. "Ten years? Twenty years? Thirty?"

"Less than five years," he said. "It's not going to be in ten years. I mean, in ten years we won't even have smartphones anymore."

*In ten years we won't even have smartphones anymore.* Whether or not he's proved right on the timeline, one thing is clear: we're headed toward a cultural shift that could be even bigger than smartphones. No more looking at another device for a bus schedule or news story that could just as easily scroll into your field of view. No more looking back and forth between that instructional video and the cutting board when you're trying to perfect your chiffonade.

That promise goes far beyond information. The VR-AR convergence promises virtual objects and people that exist in our world, as solid-looking as anything or anyone else: virtual versions of our own friends sitting in our living rooms—as themselves or as alter-ego avatars—while virtual versions of us sit in theirs. Reality itself is on the verge of becoming elastic like it's never been before. It's not like a bad acid trip, though; it's more like a conceptual Lego set with every brick you'd ever need and the ability to put them together with little more than intent. I needed someone who would help me understand what this world would look like.

I needed the guy who started it all.

## MEETING THE GODFATHER

Jaron Lanier is late, and I'm . . . well, I'm vibrating. I'm not sure whether my bouncing knee is the result of impatience, excitement, or just the umpteenth cup of coffee I've had this morning in anticipation, but I'm finding it difficult to sit still—though I hope my bladder finds it a bit easier. I'm sitting in a bar in Oakland that doubles as a cafe in the morning hours; toward the back, what looks like a group of new hires is getting trained. I picked a high-top table near the front, facing the door, just so I'd be sure to see the VR pioneer when he walked in.

Once he does show, I wonder how there could have been any doubt I'd see him. Lanier may not be hugely tall, but he cuts a memorable figure; he's stout, wearing black pants and a voluminous black T-shirt, and has a head full of long, unruly dreadlocks. Add in his high, ethereal voice, his wispy facial hair, and his heavy-lidded pale blue eyes, and it feels like he clambered out of the pages of a fantasy novel.

He only strengthens that impression when he gets going about reality—virtual, augmented, or plain old real. As the man who popularized the term "virtual reality" (Chapter 1 alert!), Lanier has long been the field's most visible trailblazer, and for more than thirty years and through numerous books has been an eloquent voice about technology and consciousness. These days, he works in Microsoft's research division, doing things he'll only describe, with a laugh, as "pretty esoteric and high-risk and maybe stuff that isn't ready for discussion." Just because he doesn't discuss his work, though, doesn't mean he's not willing to talk about where we're all headed.

And he's not necessarily an optimist. "My view of the future is that it's a contest between creativity and power seeking," he says. "If people continue to have this kind of sort of uninformed passivity in their interaction with the tech companies, there's a danger that VR could be a very, very creepy thing. The *Matrix* movies are a great example."

They really are—and they're one that people have pointed to since VR's reemergence in 2012. So I ask him what VR *should* be. He responds not with a specific vision, but with a mood. "I want people to be fine wine connoisseurs and not drunks," he says. He motions at the speakers in the bar; they're playing rock music that's loud enough to notice, even if it's not too loud to talk over. "Like, I could do without that. I actually think that that detracts from music. I'd rather come here and listen to a live band playing their hearts out; I think that's a better form of music than just leaving something on. There will probably be some VR thing like that, some ambient thing. But the real thing should be this rarified special thing you do that you really pay attention to and value—that people create with all their hearts. And by definition that can't be on all the time."

At first I want to disagree with him. I want to say that he's being needlessly precious; that to sip VR's magic is to confine ourselves to just one shade of the rainbow that presence enables. I want to, but I can't. He's right. Everything that I've experienced in VR the past five years—every emotion I've had stirred, every memory, every person whom I've met, every quickened pulse and tranquil bliss-out—has been a direct result of the fact that it was the only thing

I was experiencing at the time. That focus, that ability to sink into VR, is absolutely crucial to creating a lasting state of presence.

Interestingly, in more than fifty minutes of conversation, Lanier uses the word "presence" exactly zero times. I'm not entirely surprised; by the time the term took root in the early 1990s, he had already left VPL Research, his old company. He does, though, talk about the importance of shared experiences in VR. In fact, he says, the *original* original usage of the term "virtual reality," from the playwright Antonin Artaud, was about just that. Being with someone else in VR is "just a more interesting, cooler, more optimistic, more touching and tender and more just intense and joyous experience," Lanier says.

And there, we couldn't agree more.

# CONCLUSION

# WHEN 2020 IS HINDSIGHT
## WHAT LIFE IN 2028 MIGHT
## ACTUALLY LOOK LIKE

Bᴜᴛ ᴏɴᴇ ǫᴜᴇsᴛɪᴏɴ still remains, and it's a biggie. If all these avenues of progress and research pan out, then we're talking about much more than that ideal, lightweight pair of glasses. We're talking about a computing system that's the stuff of science-fiction legend: A little bit *Minority Report*, a little bit *Ready Player One*, a little bit *Snow Crash*, a little bit *Demolition Man*. Granted, there will certainly be unsolved elements—a haptics system that's indistinguishable from reality is far more than a decade away (to say nothing of being able to taste virtual things, because, well, *why?*)—but in 2028 we'll certainly have taken steps in that direction. What might that *look* like? What will this effortless, persistent, all-senses-on-deck system mean for your day-to-day life?

*Ugh. 6:15 already?* The sound of the alarm intrudes on your dream, but waking up isn't the struggle it once was. Ever since you got the latest generation of the LifeLenz MR, you've noticed that lucid dreaming is much more frequent: not only do you *know* when you're dreaming, but you're able to harness that knowledge to take advantage of the dream state. Goodbye sitting in your underwear in the final exam of a class you didn't even know you were enrolled in; hello flying!

You reach out in the darkness and pick your LifeLenz out of the charger, slipping the temples behind your ears. It blinks out of sleep mode just like you did, scrolling a soothing *Good morning!* across your vision. Your dog is still snoring softly, so rather than turn on the bedroom light, you whisper, "Run my wake routine." On cue, a lamp in the corner of the room glows the faint pink light of dawn, intensifying into a mellow orange. The LifeLenz's outward-facing cameras mapped all the rooms in your apartment when you first got it, but as soon as it woke up it did a quick rescan, just to make sure you hadn't rearranged your furniture or bought a new table that would change the way the light fell across the bedspread.

You walk into the bathroom and brush your teeth, your Life-Lenz feeding you only the information you need this early in the morning: a tiny icon of a sun blinks on in the upper right corner, telling you today's weather. Your calendar can wait; for now, you just want to get a sense of what it's like outside.

You slip down the hall into your study and sit on a cushion on the floor. You used to try to meditate for an hour every day, but you've found that with the new update to LenZen, your glasses' mindfulness app, ten or fifteen minutes leave you feeling as centered as ever. You look upward to activate the apps you're most likely to use at this time of day—if you don't see what you're looking for, you can always bring up a floating keyboard and tap out its name—and launch LenZen with a wink. Your glasses darken to full opacity and bring up Zion National Park at sunset. Ever since you visited there as a kid, it's been your happy place, and now, looking across the red cliffs that stretch nearly all the way around you, you can't help but smile.

You blink past LenZen's suggested meditation programs, opting for a simple mindful-breathing program with no voice cues or heart rate information; the glasses' audio quality and biometric sensors are all incredible, but you want to use just your own breathing as a guide. You gaze at the horizon as you inhale and exhale, your breath a soft blue plume that begins to tinge with yellow color if you start breathing too shallowly or quickly. After fifteen minutes that feel like two, a quiet gong sounds, and you take a moment to collect yourself before the LifeLenz returns you to your study and the rest of your day.

A quick breakfast smoothie before work; you're almost out of cinnamon, so you Lenz-scan the bar code and add it to your shopping list with a three-fingered gesture. Your LifeLenz lets you know that your bus is five minutes away, giving you plenty of time to get downstairs. You mostly just read on the bus—yes, with a physical book—but today your friend sent you a trailer for that Vmax movie you've both been waiting for, so you blink your glasses into VR mode, select which Participant Role you want to play (you usually love Reluctant Savior, but you're on the bus, so today you opt for Bystander, letting the scene play out from whatever vantage point you want) and settle in.

Work flies by, mostly research for a brief you're writing. You don't wear gloves most of the time but keep a pair at the office because they're so perfectly suited to spatial computing. You usually just project the keyboard onto your empty desk; at your last chiropractor appointment, though, your neck had been a little more kinked than usual, so you raise the virtual keyboard a few inches and tilt it toward yourself. The "keys" are as responsive as ever; you love the tactile feedback of mechanical keyboards, so you have the haptics tuned higher than most of your colleagues—thankfully, there's no audible clacking to distract them.

Every so often a co-worker in another city will ping you via video pop-up for a quick question, but you prefer to have your real meetings in person, so when it's time for your weekly brainstorm with your team, you head into a small conference room where everyone is gathered. Well, kind of; one of your five colleagues is there in flesh and blood—a "corpo," as everyone has taken to calling them—but the others are there in avatar form. All the avatars have a small *v* hovering above their heads, affirming their virtual status. After those con artists bilked an assisted-living home full of seniors out of their retirement funds in 2024, the FCC's reality regulatory committee passed a series of industry-wide mandates. Now, immersive computing devices auto-tag any avatar or other artificial entity that's rated a 6 or above on the VirTuring Scale (meaning that it possesses either visual fidelity or AI complexity indistinguishable from the real thing). Properly zoned entertainment centers can submit for exemptions, of course. After all, why would you visit a Frightrift, or a *Flesh*rift, if you were constantly going to be reminded of the illusion?

After work, you run along the waterfront. The weather's been beautiful lately, so you want to make the most of it. You stop to stretch halfway through and notice another runner doing the same thing—and, wait, he's cute. He looks up at the same time and notices you. If it were passing eye contact, nothing would happen, but holding each other's gaze for just a moment engages your glasses' Social Mode, showing you the information about the other person they're broadcasting. People configure their privacy settings to their own liking, but "Passerby" by default broadcasts little more than one's mood; you can find out more only if, Tinder-style, you both choose to bump the other up a level to "Acquaintance." (Don't worry, Tinder's still around; it's just mostly bots at this point. Which is fine! Every now and then, people want to spend a little time with an AI avatar who knows everything they like . . . and has a ridiculous body to boot.)

A quick drink with two old friends from college—one avatar, one corpo—and you head home for the night. You're looking forward to a little me time; last night was your weekly multiplayer gaming session, and your living room was a literal battlefield until almost midnight. That new Playsuit you picked up during a closeout sale might have been last year's model, but it was responsive enough that when a teammate in Dallas bear-hugged you, you felt it.

Tonight, though, is relaxation. You pour a glass of wine, settle back on the couch, click your glasses into full VR, and launch Omninote, a live-music experience. Your favorite band is playing a three-night engagement, and you snagged a great seat. *Everyone* got great sets, in fact; that's the magic of Omninote. You minimize all the optional visual effects—you want to just listen tonight rather than see the real-time hallucination spun by a "reality performance artist."

A few songs in, you feel a faint buzz. You nod your head slightly, overriding the "do not disturb" setting, and lowercase italic text slides in from the left: *enjoying the show?* It's the cute runner—he must be here too. Good thing you left your default avatar on; in your party getup, you might have been unrecognizable. You sit up a bit, looking around the small club. *There* he is. You smile. So does he. He gestures toward the bar in the back, and you see his mouth forming the same words sliding into your vision: *want a drink?*

Of course you do. There's a moment in the making.

# ACKNOWLEDGMENTS

Fun fact: the acknowledgments section is my favorite part of any book. Some are dry, some are disarming, some are pompous—okay, a lot are pompous—but they make a book feel *alive,* like some sort of complex organism that only exists thanks to this magical concert of disparate relationships.

Writing one, though? Not as easy as you'd think. For one, I have no idea how far back I'm supposed to go. Obviously, any chance I had at stringing together this many words in a row starts with my parents. Being the kid of a research librarian (mom) and a college professor (dad) means that I grew up in a house where books and ideas and words *mattered*—wordplay too, thankfully—and while I absolutely despised writing until sometime in my late teens, there's not a chance that I would have considered it without them.

But after that, where do I go? The teachers who valued critical thought over rote curricula? Sally Harvey, Bob Courtney, Carla Gardner, Greg Mongold, Craig Wilder, Phyllis Garland? Sam Freedman, who finally told me to stop using so many words and just write? (And who also did me the favor of telling me I wasn't ready for his book-writing course in grad school—because that would have been a disaster for everyone involved.)

And once I actually figured out I wanted to write, who then? The writers and editors at Art Cooper's early-'00s *GQ*, where I learned from a murderer's row of incredible talents? Devin Friedman, Brandon Holley, Adam Sachs, Lucy Kaylin, Andrew Corsello, Chris Raymond, Jim Nelson, Adam Rapoport, Michael Hainey, Marty Beiser, Mark Healy—even writing those names out makes me feel like a twenty-four-year-old freelance fact-checker again, but every word I wrote there was at its root emulation, trying to mimic my way into some umbral approximation of their work.

Later, at *Complex*, I got the chance to learn all over again, working alongside people who became family as we weathered the upheaval of a recession and learned how to pivot: Noah Callahan-Bever, Donnie Kwak, Justin Monroe, Anoma Ya Whittaker, Tim Leong, Jack Erwin, Damien Scott, Joe La Puma, Bradley Carbone, Mary H. K. Choi.

And at *WIRED*, where after my first week I said the same thing everyone says there after their first week ("holy shit, this is the smartest group of people I've ever been around"), I learned not just how to be a better magazine editor and writer, but a better storyteller in all capacities. There are way, *way* too many people there to thank after more than six years there, but I'll try to limit it to the people that I've worked with on VR-related stuff: Scott Dadich, Rob Capps, Jason Tanz, Caitlin Roper, Jason Kehe, Angela Watercutter, Sarah Fallon, Jon Eilenberg, Adam Rogers. That's just a start, though; there are so, so many other incredible editors and writers I'm lucky to call not just colleagues but friends. And to the research, copy, design, photography, production, social, video, and every other department, past and present: you're all amazing. I wish I could thank you all by name.

(Side note: Man, this really *is* hard. What's taking me so long?)

Formative experiences aside, this book wouldn't exist without my inimitable editor Hilary Lawson, who not only emailed me out of the blue because she'd read an essay I'd written and thought that I might have a book in me, but who managed to keep me sane while I took *way* too much time writing this book. ("But the longer I wait, the more stuff happens!" I told her no fewer than three times over coffee, while she managed to make her patient grimace look like a smile.) To the unflappable champions at HarperOne who made this book better in innumerable ways—Sydney Rogers, Lisa Zuniga, Jessie Dolch, Ann Edwards, Melinda Mullin, Courtney Nobile, and everyone else—thank you. And Tiffany Kelly, who managed to fact-check this book in record time: I'm still looking for a third confirmation of how to spell "gratitude," but believe me, you've got mine.

Thank you, Scrivener, for being an amazing writing tool that somehow harnessed a zillion interview transcripts, research PDFs, pages of handwritten notes, and assorted disjointed ramblings into something resembling coherence. Thank you to everyone in the VR world who was willing to share their expertise and genius with me, both on and off the record. Thank you to the users—the passionate, curious, invested early adopters who support VR's promise not just with their purchasing dollars but with their hearts and minds. The future of presence has you to thank.

But most of all, I wouldn't have a future without my wife, partner, and best friend—so thank you, Kelli. Thank you for believing in me, thank you for trusting me, and thank you for talking me off the writer-anxiety ledge more times than I can count. Sitting here on the couch writing this right now, I've got you on one side and a tiny sleeping panda-bat-looking dog on the other (hi, Crosby!). Life outside the headset really couldn't be any better.

Wait, though; there's still one left. So thank *you,* dear person who read all the way to this point without throwing the book away. There wouldn't be an acknowledgments section without you. I'll see you in the Metaverse, I hope.

# REFERENCES

## INTRODUCTION: WELCOME TO VIRTUAL REALITY

*1* When the video starts . . . Paul Rivot, "My 90 year old grandmother tries the Oculus Rift," YouTube, April 14, 2013, https://www.youtube.com /watch?v=pAC5SeNH8jw.

*3* VR helps soldiers find relief from post-traumatic stress disorder . . . Adam Popescu, "These VR Systems Help Treat Veterans Recovering from PTSD," *Bloomberg Businessweek*, March 20, 2017, https://www.bloomberg.com /news/articles/2017-03-16/these-vr-systems-help-treat-veterans-recovering -from-ptsd.

*3* may one day help it lessen the opioid epidemic . . . Megan Molteni, "Opioids Haven't Solved Chronic Pain. Maybe Virtual Reality Can," *WIRED*, November 2, 2017, https://www.wired.com/story/opioids-havent -solved-chronic-pain-maybe-virtual-reality-can/.

*3* Real estate companies . . . John Gaudiosi, "Now you can shop for luxury homes in virtual reality," *Fortune*, September 9, 2015, http://fortune.com /2015/09/09/virtual-reality-real-estate/.

*4* to examine a 3-D scan of the baby's heart using VR . . . Gillian Mohney, "How a Google Cardboard Virtual Reality Device Helped a Surgeon Operate on Infant's Heart," ABC News, January 8, 2016, http://abcnews .go.com/Health/google-cardboard-virtual-reality-device-helped-surgeon -operate/story?id=36167944.

4 causing a bead of sweat to trickle down your back . . . Harvard Health Center, "Understanding the Stress Response," March 2011, updated March 18, 2016, https://www.health.harvard.edu/staying-healthy /understanding-the-stress-response.

4 hearing a choir gives you chills . . . Valorie N. Salimpoor et al., "Anatomically Distinct Dopamine Release During Anticipation and Experience of Peak Emotion to Music," *Nature Neuroscience* 14 (2011): 257–62, https:// doi.org/10.1038/nn.2726.

5 Social media filled with . . . Jake Silverstein, "Editor's Letter: Take Flight With Virtual Reality," *New York Times*, December 10, 2015, https://www .nytimes.com/2015/12/13/magazine/editors-letter-take-flight-with -virtual-reality.html.

## CHAPTER 1: PRESENCE

20 When Ivan Sutherland was a student at MIT . . . Harry McCracken, "A Talk with Computer Graphics Pioneer Ivan Sutherland," *TIME*, April 12, 2013, http://techland.time.com/2013/04/12/a-talk-with-computer -graphics-pioneer-ivan-sutherland/.

21 Artaud first described theater as "virtual reality" . . . Antonin Artaud, "The Alchemical Theater," in *The Theater and Its Double* (New York: Grove, 1988), 48–52.

21 a program the Air Force called Super Cockpit . . . Thomas A. Furness III, "The Super Cockpit and Its Human Factors Challenges," *Proceedings of the Human Factors Society Annual Meeting* 30 (1986): 48–52, https://doi .org/10.1177/154193128603000112.

22 Imagine having the power to instantly change your environment . . . "NASA's Virtual Workstation," *NASA Tech Briefs*, July/August 1988, 20–21.

23 over the protests of his colleagues . . . Jaron Lanier, "Virtual Reality: A Techno-Metaphor With a Life of Its Own," *The Whole Earth Catalog*, Fall 1999, 16–18.

23 All that fantasy came at a price . . . A.J.S. Rayl, "The New, Improved Reality: Will the Ultimate Connection Between Humans and Computers Become the Ultimate Escape?" *Los Angeles Times*, July 21, 1991, http:// articles.latimes.com/1991-07-21/magazine/tm-264_1_virtual-reality/2.

23 VPL would ultimately file for bankruptcy in the 1990s . . . Frank Steinicke, *Being Really Virtual: Immersive Natives and the Future of Virtual Reality* (Switzerland: Springer International Publishing, 2016), 14–15.

24 King successfully sued . . . Casey Davidson, "Stephen King Wins Lawsuit," *Entertainment Weekly*, April 22, 1994, http://www.ew.com/article/1994 /04/22/stephen-king-wins-lawsuit.

24 Nintendo tried to capitalize . . . Benj Edwards, "Unraveling the Enigma of Nintendo's Virtual Boy, 20 Years Later," *Fast Company*, August 21, 2015, https://www.fastcompany.com/3050016/unraveling-the-enigma-of -nintendos-virtual-boy-20-years-later.

25 the same year the first photo . . . Andrew Hough, "How the First Photo Was Posted on the Web 20 Years Ago," *The Telegraph*, July 11, 2012, http:// www.telegraph.co.uk/technology/news/9391110/How-the-first-photo -was-posted-on-the-Web-20-years-ago.html.

30 more than two thousand pixels *per inch* . . . Steven M. LaValle, *Virtual Reality* (Cambridge: Cambridge University Press, 2017), 140.

31 exactly where your pupils are focused . . . Peter Rubin, "Has This Stealth Company Solved Vision-Quality VR?," *WIRED*, June 19, 2017, https://www .wired.com/story/varjo-vr-microdisplay/.

31 as if the technology was not involved in the experience . . . "Presence Defined," International Society for Presence Research, https://ispr.info /about-presence-2/about-presence/.

## CHAPTER 2: ALONE ON A MOUNTAINTOP

44 more than three-quarters of all people . . . American Psychological Association, American Institute of Stress, NY, "Stress Research," Stress.org, July 8, 2014, https://www.stress.org/stress-research/.

44 After the popularity of meditation and yoga . . . Holly Hammond, "Yoga Pioneers: How Yoga Came to America," *Yoga Journal*, August 29, 2007, https://www.yogajournal.com/yoga-101/yogas-trip-america.

44 By the mid-'90s . . . Virginia Heffernan, "The Muddied Meaning of 'Mindfulness,'" *New York Times*, April 14, 2015, https://www.nytimes.com/2015 /04/19/magazine/the-muddied-meaning-of-mindfulness.html.

45 Chopra and his rumored $80 million net worth . . . Peter Rowe, "Truly, madly, deeply Deepak Chopra," *San Diego Union-Tribune*, May 3, 2014, http://www.sandiegouniontribune.com/lifestyle/people/sdut-truly -madly-deepak-chopra-2014may03-htmlstory.html.

45 Various studies have found that meditation and mindfulness training . . . Judson A. Brewer et al., "Meditation Experience Is Associated with Differences in Default Mode Network Activity and Connectivity," *Proceedings of the National Academy of Sciences USA* 108, no. 50 (2011): 20254–59, https:// doi.org/10.1073/pnas.1112029108.

45 a study conducted at a Harvard Medical School–affiliated hospital . . . Britta K. Hölzel et al., "Mindfulness Practice Leads to Increases in Regional Brain Gray Matter Density," *Psychiatry Research* 191, no. 1 (2011): 36–43, http://doi.org/10.1016/j.pscychresns.2010.08.006.

45 Google began offering its employees meditation classes . . . Noah Shachtman, "In Silicon Valley, Meditation Is No Fad. It Could Make Your Career," *WIRED*, June 2013, https://www.wired.com/2013/06/meditation -mindfulness-silicon-valley/.

48 76 percent of people who underwent the experience . . . Carrie Heeter, "*Being* There: The Subjective Experience of Presence," *Presence: Teleoperators and Virtual Environments* 1, no. 2 (1992): 262–71, https://doi.org/10 .1162/pres.1992.1.2.262.

48 Even when we're relaxing, our brains are as active . . . Louis Sokoloff et al., "The Effect of Mental Arithmetic on Cerebral Circulation and Metabolism," *Journal of Clinical Investigation* 34, no. 7, pt. 1 (1955): 1101–08, https://doi.org/10.1172/JCI103159.

49 The default-mode network's chief purpose . . . Randy L. Buckner, Jessica R. Andrews-Hanna, and Daniel L. Schacter, "The Brain's Default Network: Anatomy, Function, and Relevance to Disease," *Annals of the New York Academy of Sciences* 1124 (2008): 1–38, https://doi.org/10.1196/annals .1440.011.

51 mindfulness meditation seems to heighten interoception . . . Norman A. S. Farb, Zindel V. Segal, and Adam K. Anderson, "Mindfulness Meditation Training Alters Cortical Representations of Interoceptive Attention," *Social Cognitive and Affective Neuroscience* 8, no. 1 (2013): 15–26, https:// doi.org/10.1093/scan/nss066.

55 he became one of Twitter's first employees . . . Jessica Guynn and Chris

O'Brien, "In San Francisco, Twitter's IPO stokes debate over wealth it created," *Los Angeles Times,* November 11, 2103, http://articles.latimes .com/2013/nov/11/business/la-fi-twitter-boom-20131112.

56   a new social system that blends fantasy and reality . . . "Virtual Reality, Real Benefits," Virtual Reality Pop, June 6, 2016, https://virtualrealitypop .com/virtual-reality-real-benefits-1eb15625f8b8#.nehv8hm6h.


## CHAPTER 3: HEDGEHOG LOVE

65   offers further evidence that you exist . . . Carrie Heeter, *"Being* There."

65   Social presence can also be created . . . Heeter, *"Being* There."

66   Since Oliver Hardy first shot an exasperated look . . . Steve Seidman, "Performance, Enunciation and Self-Reference in Hollywood Comedian Comedy," in *Hollywood Comedians: The Film Reader,* ed. Frank Krutnik (London: Routledge, 2003), 24.

66   but a *result* of it . . . Joan Kellerman, James Lewis, and James Laird, "Looking and Loving: The Effects of Mutual Gaze on Feelings of Romantic Love," *Journal of Research in Personality* 23 (1989): 145–61, https://doi.org /10.1016/0092-6566(89)90020-2.

70   However, Oculus Story Studio is no more . . . Jason Rubin, "The Next Chapter of Creative Development in VR," Oculus, May 4, 2017, https:// www.oculus.com/blog/the-next-chapter-of-creative-development-in-vr/.

71   1895's *Arrival of a Train at La Ciotat* . . . Ray Zone, *Stereoscopic Cinema and the Origins of 3-D Film, 1838–1952* (Lexington: University Press of Kentucky, 2007), 76.

71   1902's *A Trip to the Moon* . . . Tim Dirks, "Filmsite Movie Review: *Voyage dans la Lune* (A Trip to the Moon) (1902)," Filmsite, accessed January 20, 2018, http://www.filmsite.org/voya.html.

71   In *Invasion!,* for example . . . Andy Eddy, "Baobab Studios Earnestly Starts Its 'Invasion!' of the VR World," UploadVR, August 7, 2016, https:// uploadvr.com/baobab-studios-earnestly-starts-invasion-vr-world/.

75   in one study, volunteers who were given . . . Ivelina V. Piryankova et al., "Owning an Overweight or Underweight Body: Distinguishing the Physical, Experienced and Virtual Body," *PLOS ONE* 9, no. 8 (2014): e103428, https://doi.org/10.1371/journal.pone.0103428.

76 by closing and opening your hands . . . Sam Byford, "Valve Shows Off New VR Controller Prototypes," The Verge, October 12, 2016, http://www.theverge.com/2016/10/12/13264950/valve-vive-vr-controllers-new-prototype.

77 embedded in any smartphone or VR headset . . . Jamie Feltham, "Leap Motion Showcases Mobile Hand-Tracking in New Video," UploadVR, March 15, 2017, https://uploadvr.com/leap-motion-showcases-mobile-hand-tracking-new-video/.

77 HTC sells small wearable trackers . . . "Vive Tracker: Go Beyond VR Controllers," Vive, https://www.vive.com/us/vive-tracker/.

77 wearing a headset and a huge grin on his face . . . Mark Zuckerberg, Facebook, February 9, 2017, https://www.facebook.com/photo.php?fbid=10103490859471431.

79 When it premiered at the Cannes Film Festival . . . Pete Hammond, "Academy Votes Special Oscar to Alejandro G. Inarritu's Virtual Reality Installation 'Carne y Arena,'" Deadline, October 27, 2017, http://deadline.com/2017/10/academy-votes-special-oscar-to-alejandro-g-inarritus-virtual-reality-installation-carne-y-arena-1202196302/.

81 If you give people too much ability . . . Adi Robertson, "Allumette Is a Beautiful Virtual World from Oculus Story Studio Veteran," The Verge, January 26, 2016, http://www.theverge.com/2016/1/26/10831946/allumette-virtual-reality-animated-film-penrose-sundance-2016.

## CHAPTER 4: EMPATHY VS. INTIMACY

85 let's see what the Oxford Dictionary of Sociology . . . John Scott and Gordon Marshall, eds., A Dictionary of Sociology, 3rd rev. ed. s. v. "empathy," "intimacy" (Oxford: Oxford University Press, 2009), 214, 373–74, https://doi.org/10.1093/acref/9780199533008.001.0001.

88 In 2007, she used the virtual community . . . "Gone Gitmo—A Virtual Guantanamo Bay Prison Built in Second Life," Immersive Journalism, May 12, 2010, http://www.immersivejournalism.com/gone-gitmo/.

88 it was like the video game The Sims . . . Kristin Kalning, "If Second Life Isn't a Game, What Is It?," NBC News, updated March 12, 2007, http://www.nbcnews.com/id/17538999/ns/technology_and_science-games/t/if-second-life-isnt-game-what-it/#.WRDyl4krLdR.

88 nearly a million people still use it every month . . . Emanuel Maiberg, "Why Is 'Second Life' Still a Thing?," Motherboard, April 29, 2016, https:// motherboard.vice.com/en_us/article/why-is-second-life-still-a-thing -gaming-virtual-reality.

88 Chris Milk himself among them . . . Caleb Garling, "Virtual Reality, Empathy and the Next Journalism," *WIRED*, https://www.wired.com /brandlab/2015/11/nonny-de-la-pena-virtual-reality-empathy-and-the -next-journalism/.

88 In one video viewable on YouTube . . . Nonny de la Peña, "GinaRodriguez Cries.MOV," YouTube, January 22, 2012, https://www.youtube.com /watch?v=kbUmc9BRHgM.

93 misters and hypnotists . . . Adam Epstein, "Hollywood's Fantastic, Failed Attempts to Make Audiences Smell and 'Feel' Movies, from AromaRama to 4D," Quartz, April 1, 2016, https://qz.com/649920/hollywoods-fantastic -failed-attempts-to-make-audiences-smell-and-feel-movies-from -aromarama-to-4d/.

93 more and more of our media consumption is happening on our phones . . . Amy Mitchell et al., "1. Pathways to News," Pew Research Center, July 7, 2016, http://www.journalism.org/2016/07/07/pathways- to-news/ (news); Matthew Ingram, "The Smartphone Is Eating the Television, Nielsen Admits," *Fortune*, December 7, 2015, http://fortune .com/2015/12/07/smartphone-tv-report/ (television); Lee Rainie and Andrew Perrin, "Slightly Fewer Americans Are Reading Print Books, New Survey Finds," Pew Research Center, October 19, 2015, http://www .pewresearch.org/fact-tank/2015/10/19/slightly-fewer-americans-are -reading-print-books-new-survey-finds/ (books).

94 he wrote in *The Atlantic* . . . Paul Bloom, "It's Ridiculous to Use Virtual Reality to Empathize with Refugees," *The Atlantic*, February 3, 2017, https://www.theatlantic.com/technology/archive/2017/02/virtual-reality -wont-make-you-more-empathetic/515511/.

94 twice what was expected . . . Angela Watercutter, "VR Films Work Great for Charity. What About Changing Minds?," *WIRED*, March 1, 2016, https://www.wired.com/2016/03/virtual-reality-social-change- fundraising/; "Syrian Refugee Crisis," UNVR, United Nations Virtual Reality, https://unitednationsvirtualreality.wordpress.com/virtual -reality/cloudsoversidra/.

*96*  where chip manufacturers cluster . . . "Don Hoefler, Writer Who Coined the Term 'Silicon Valley,'" *San Jose Mercury News*, April 16, 1986.

*99*  a doctoral dissertation on the topic of lightfield capture . . . Ren Ng, "Digital Light Field Photography," PhD diss., Stanford University, June 2006.

*100*  allowing for movement within the environment . . . Angela Watercutter, "The Incredible, Urgent Power of Remembering the Holocaust in VR," *WIRED*, April 20, 2017, https://www.wired.com/2017/04/vr-holocaust -history-preservation/.

*100*  for more indie-film-friendly prices . . . Ken Yeung, "Facebook Takes on Lytro with New Surround 360 Cameras That Shoot in 6 Degrees of Freedom," Venture Beat, April 19, 2017, https://venturebeat.com/2017/04 /19/facebook-takes-on-lytro-with-new-surround-360-cameras-that -shoot-in-6-degrees-of-freedom/.

## CHAPTER 5: WHAT TO DO AND WHO TO DO IT WITH

*113*  If you had to choose between muffins . . . Devon Ivie, "Every Poignantly Weird Question Reggie Watts Has Asked Guests on *The Late Late Show*," Vulture, February 23, 2017, http://www.vulture.com/2015/10/reggie-watts -late-late-show-questions.html.

*114*  Altspace claims it can handle . . . "Breakthrough Tech Gives Everyone a Front Row Seat at Events," AltspaceVR, May 20, 2016, https://altvr.com /breakthrough-tech-gives-everyone-a-front-row-seat-at-events/.

*117*  how being in VR affected people's concept of personal space . . . Jeremy N. Bailenson et al., "Equilibrium Theory Revisited: Mutual Gaze and Personal Space in Virtual Environments," *Presence: Teleoperators and Virtual Environments* 10, no. 6 (2001): 583–98, https://doi.org/10.1162 /105474601753272844.

*119*  even rape threats and other violent overtures . . . Laura Hudson, "Curbing Online Abuse Isn't Impossible. Here's Where We Start," *WIRED*, May 15, 2014, https://www.wired.com/2014/05/fighting-online-harassment/; Amy O'Leary, "In Virtual Play, Sex Harassment Is All Too Real," *New York Times*, August 1, 2012, http://www.nytimes.com/2012/08/02/us/sexual -harassment-in-online-gaming-stirs-anger.html.

*120*  toward my virtual crotch and began rubbing . . . Jordan Belamire, "My First Virtual Reality Groping," Athena Talks, October 20, 2016, https://

medium.com/athena-talks/my-first-virtual-reality-sexual-assault
-2330410b62ee.

120 apologizing for their failure to anticipate the episode . . . Henry Jackson,
"Dealing with Harassment in VR," UploadVR, October 25, 2016, https://
uploadvr.com/dealing-with-harassment-in-vr/.

121 another user would simply disappear . . . Jamie Feltham, "'BigScreen'
Adds a Personal Space Bubble to Stop VR Trolls in Their Tracks,"
UploadVR, June 13, 2016, https://uploadvr.com/bigscreen-adds-personal
-space-bubble-stop-vr-trolls-tracks/.

121 during her first time on the platform . . . Taylor Lorenz, "Virtual Reality
Is Full of Assholes Who Sexually Harass Me. Here's Why I Keep Going
Back," Mic, May 26, 2016, https://mic.com/articles/144470/sexual
-harassment-in-virtual-reality#.ne3AlqaPQ.

121 available inside VR every minute of the day . . . "Introducing Space Bubble," AltspaceVR, July 13, 2016, https://altvr.com/introducing-space
-bubble/.

122 cut down on harassment in its comments . . . Nicholas Thompson, "Mr.
Nice Guy," WIRED (September 2017): 82–89, https://www.wired.com
/2017/08 /instagram-kevin-systrom-wants-to-clean-up-the-internet/.

122 silent participation rather than active engagement . . . J. Fox and W. Y.
Tang, "Sexism in Online Video Games: The Role of Conformity to Mas-
culine Norms and Social Dominance Orientation," Computers in Human
Behavior 33 (2014): 314–20, https://doi.org/10.1016/j.chb.2013.07.014.

123 affected people's mental states in the real world . . . Oswald D. Kothgassner
et al., "Real-Life Prosocial Behavior Decreases After Being Socially
Excluded by Avatars, Not Agents," Computers in Human Behavior 70
(2017): 261–69, https://doi.org/10.1016/j.chb.2016.12.059.

124 just as alienation can be induced, so can altruism . . . Robin S. Rosenberg,
Shawnee L. Baughman, and Jeremy N. Bailenson, "Virtual Superheroes:
Using Superpowers in Virtual Reality to Encourage Prosocial Behavior,"
PLOS ONE 8, no. 1 (2013): e55003, https://doi.org/10.1371/journal.pone
.0055003.

126 the company's founder wrote in a post . . . Darshan Shankar, "$3 Million
Financing Round Led by Andreessen Horowitz," Bigscreen VR, February 24,
2017, https://blog.bigscreenvr.com/3-million-financing-round-led-by
-andreessen-horowitz-1a535c8c36a8.

*126*  Rosedale left the company in 2009 . . . Rowland Manthorpe, "Remember Second Life? Now It's Being Reborn in Virtual Reality," *Wired UK*, October 24, 2016, http://www.wired.co.uk/article/philip-rosedale-high-fidelity.

*126*  ones that users can enable with extra sensors . . . Tom Simonite, "The Quest to Put More Reality in Virtual Reality," *MIT Technology Review*, October 22, 2014, https://www.technologyreview.com/s/531751/the-quest -to-put-more-reality-in-virtual-reality/.

## CHAPTER 6: THE STARRY NIGHT THAT WASN'T THERE

*130*  entire experiences and adventures . . . Mark Zuckerberg, "I'm excited to announce that we've agreed to acquire Oculus VR, the leader in virtual reality technology," Facebook, March 25, 2014, https://www.facebook .com/zuck/posts/10101319050523971.

*130*  a team dedicated to building "VR social apps" . . . "Mark Zuckerberg MWC 2016 Keynote," YouTube, Great Presentation, March 12, 2016, https://www.youtube.com/watch?v=TKP55_2EKRQ.

*131*  then things took a hands-on turn . . . Jessi Hempel, "Facebook Believes Messenger Will Anchor a Post-App Internet," *WIRED*, April 12, 2016, https://www.wired.com/2016/04/facebook-believes-messenger-will -anchor-post-app-internet/.

*139*  Memories can be divided into two major types . . . Neil W. Mulligan, "Memory: Implicit versus Explicit," in Encyclopedia of Cognitive Science (John Wiley & Sons, Ltd., 2006), https://doi.org/10.1002/0470018860 .s00572.

*140*  codified the idea of "laboratory memory" . . . Roberto Cabeza et al., "Brain Activity During Episodic Retrieval of Autobiographical and Laboratory Events: An fMRI Study Using a Novel Photo Paradigm," *Journal of Cognitive Neuroscience* 16, no. 9 (2006): 1583–94, https://doi.org/10.1162 /0898929042568578.

*140*  the adoption of VR in experimental psychology . . . Benjamin Schöne, Marlene Wessels, and Thomas Gruber, "Experiences in Virtual Reality: A Window to Autobiographical Memory," *Current Psychology* (2017): 1–5, https://doi.org/10.1007/s12144-017-9648-y.

*142*  helps people navigate more efficiently . . . Stuart C. Grant and Lochlan E. Magee, "Contributions of Proprioception to Navigation in Virtual Envi-

ronments," *Human Factors* 40, no. 3 (1998): 489–97, https://doi.org/10.1518
/001872098779591296.

142   helps them create a better mental map of a space . . . Roy A. Ruddle and
Simon Lessels, "The Benefits of Using a Walking Interface to Navigate Virtual Environments," *ACM Transactions on Computer-Human Interaction*
16, no. 1 (2009): 1–18, https://doi.org/10.1145/1502800.1502805.

142   your memory of the surroundings suffers . . . Najate Jebara et al., "Effects
of Enactment in Episodic Memory: A Pilot Virtual Reality Study with
Young and Elderly Adults," *Frontiers in Aging Neuroscience* 6, no. 338
(2014), https://doi.org/10.3389/fnagi.2014.00338.

142   become less mobile as they age . . . Claudia Repetto et al., "Virtual
Reality As an Embodied Tool to Enhance Episodic Memory in Elderly,"
*Frontiers in Psychology* 7, no. 1839 (2016), https://doi.org/10.3389/fpsyg
.2016.01839.

145   the effects of streaming delays on consumers . . . "The Stress of Streaming
Delays," *Ericsson Mobility Report,* February 2016, https://www.ericsson
.com/assets/local/mobility-report/documents/2016/ericsson-mobility
-report-feb-2016-interim.pdf.

145   train car with art deco touches . . . LuisMiguel Samperio, "Neurons Inc.
Biometric VR Experiment," YouTube, May 28, 2017, https://www.youtube
.com/watch?v=vR6XLknYUhA.

146   more effective than real-life exposure therapy . . . M. B. Powers and P. M. G.
Emmelkamp, "Virtual Reality Exposure Therapy for Anxiety Disorders: A
Meta-Analysis," *Journal of Anxiety Disorders* 22 (2008): 561–69, https://doi
.org/10.1016/j.janxdis.2007.04.006.

## CHAPTER 7: REC ROOM CONFIDENTIAL

150   a "virtual reality social club" . . . "About," Against Gravity, accessed February 9, 2017, https://www.againstgrav.com/press/.

152   a Japanese roboticist named Masahiro Mori . . . Lisa Katayama, "How
Robotics Master Masahiro Mori Dreamed Up the 'Uncanny Valley,'"
*WIRED*, November 29, 2011, https://www.wired.com/2011/11/pl_mori/.

152   and the hand becomes uncanny . . . Masahiro Mori, "The Uncanny Valley:
The Original Essay by Masahiro Mori," IEEE Spectrum, June 12, 2012,

https://spectrum.ieee.org/automaton/robotics/humanoids/the-uncanny
-valley.

153 Facebook's Mike Booth has said . . . Oculus, "Social VR: A Conversation
with Mike Booth," YouTube, October 12, 2016, https://www.youtube.com
/watch?v=0EZn50XCueI.

154 only about twenty fundamental facial expressions . . . Shichuan Du, Yong
Tao, and Aleix M. Martinez, "Compound Facial Expressions of Emotion,"
*Proceedings of the National Academy of Sciences USA* 111, no. 15 (2014):
E1454–62, https://doi.org/10.1073/pnas.1322355111.

154 more positive words afterward to describe the experience . . . Soo Youn
Oh et al., "Let the Avatar Brighten Your Smile: Effects of Enhancing
Facial Expressions in Virtual Environments," *PLOS ONE* 11, no. 9 (2016):
e0161794, https://doi.org/10.1371/journal.pone.0161794.

155 facial expressions to drive those of your VR avatar . . . Justus Thies et al.,
"FaceVR: Real-Time Facial Reenactment and Eye Gaze Control in Virtual
Reality," arXiv, October 11, 2016, https://arxiv.org/abs/1610.03151v1;
Veeso, "Veeso—The first Face Tracking VR Headset," YouTube, July 20,
2016, https://www.youtube.com/watch?v=xMgoypPBEgw.

158 a third of all marriages began online . . . John T. Cacioppo et al., "Marital
Satisfaction and Break-Ups Differ Across On-Line and Off-Line Meeting
Venues," *Proceedings of the National Academy of Sciences USA* 110, no. 25
(2013): 10135–40, https://doi.org/10.1073/pnas.1222447110.

163 The first time was back in 1994 . . . "Wedding in Cyberspace," Bloomberg
News, August 22, 1994, https://www.bloomberg.com/news/articles
/1994-08-21/wedding-in-cyberspace.

## CHAPTER 8: REACH OUT AND TOUCH SOMEONE

166 transformed their garage into a maze of horrors . . . Michael McFall, "The
Ambitious Dream of Evermore Park on Its Way to Reality in Utah," *Salt
Lake Tribune*, September 5, 2014, http://archive.sltrib.com/article.php?id
=58306650&itype=cmsid.

166 ultimately expanding to occupy a half acre of land . . . "The Making of
a Dream, Evermore Park and The VOID," The Void, August 25, 2016,
https://blog.thevoid.com/the-making-of-a-dream/.

*166*  in 2014 he unveiled his plan . . . Gephardt Daily, "Evermore Park Teases Us at Comic-Con 2014," YouTube, September 22, 2014, https://www.youtube .com/watch?v=AJ0Cg7UCUxc.

*166*  to adapt to holidays like Halloween and Christmas . . . Josh Young, "Evermore to Be First Adventure Park Opening in 2015," Theme Park University, June 2, 2014, http://themeparkuniversity.com/theme-parks-101 /evermore-first-adventure-park-opening-2015/.

*166*  an idea he'd wanted to realize . . . "The VOID Is Leading Virtual Reality to Victory," The Void, July 20, 2016, https://blog.thevoid.com/void-helping -people-embrace-vr/.

*166*  a scuttled movie adaptation of Little Red Riding Hood . . . Klassicalmuzik, "Red Riding Hood (TAG Entertainment) Movie Trailer (2002)," YouTube, December 1, 2015, https://www.youtube.com/watch?v=OWpS1bcxnlc.

*166*  proper placement in the CGI world . . . Sandman Studios Entertainment, "Red Lighthouse Exterior 3D Model Animation Composite, YouTube, September 23, 2014, https://www.youtube.com/watch?v=8tnw5dzhUqk.

*169*  49 percent more than you *think* you've rotated . . . Frank Steinicke et al., "Estimation of Detection Thresholds for Redirected Walking Techniques," *IEEE Transactions on Visualization and Computer Graphics* 16, no. 1 (2010): 17–27, https://doi.org/10.1109/TVCG.2009.62.

*171*  face to face from body to body . . . Shanyang Zhao, "Toward a Taxonomy of Copresence," *Presence: Teleoperators and Virtual Environments* 12, no. 5 (2003): 445–55, https://doi.org/10.1162/105474603322761261.

*173*  made out of micropumps and acoustic devices . . . Kazuki Hashimoto and Nakamoto Takamichi, "Tiny Olfactory Display Using Surface Acoustic Wave Device and Micropumps for Wearable Applications," *IEEE Sensors Journal* 16, no. 12 (2016): 4874–80, https://doi.org/10.1109/JSEN.2016.2550486.

*173*  electrostimulate the tongue to induce taste sensations . . . Satoru Sakurai et al., "Mechanism of inhibitory effect of cathodal current tongue stimulation on five basic tastes," in 2016 IEEE Virtual Reality Conference (Los Alamitos, CA: IEEE, 2016), 279–80, https://doi.org/10.1109/VR.2016.7504762.

*173*  the feelings of ostracization that can arise . . . Mariana von Mohr, Louise P. Kirsch, and Aikaterini Fotopoulou, "The Soothing Function of Touch: Affective Touch Reduces Feelings of Social Exclusion," *Scientific Reports* 7, no. 13516 (2017), https://doi.org/10.1038/s41598-017-13355-7.

*174*  evolved from silent movies a few years earlier . . . Laura Frost, "Huxley's Feelies: The Cinema of Sensation in *Brave New World*," *Twentieth Century Literature* 52, no. 4 (2006): 443–73, https://doi.org/10.1215/0041462X -2006-1001.

*174*  the book's memorable (and, okay, straight-up offensive) feely scene . . . Aldous Huxley, *Brave New World* (New York: Harper Perennial, [1932] 2006), 167–68.

*175*  made your motel bed vibrate for a quarter . . . Tom Sowa, "Magic Fingers Still Alive, But Nearing Its Finale," *Spokane Spokesman-Review*, October 16, 2012, http://www.spokesman.com/stories/2012/oct/16/aye-theres-the-rub/.

*175*  began to happen in the 1960s in research labs . . . J. K. Salisbury and Mandayam A. Srinivasan, "Phantom-Based Haptic Interaction with Virtual Objects," *IEEE Computer Graphics and Applications* 17, no. 5 (1997): 6–10, https://doi.org/10.1109/MCG.1997.1626171.

*175*  could justifiably seem to be a far-out idea . . . A. Michael Noll, "Man-Machine Tactile Communication," *SID Journal* 1, no. 2 (1972), 5–11, http:// noll.uscannenberg.org/PDFpapers/SID%20Tactile.pdf.

*176*  the office of President Richard Nixon's science advisor . . . A. Michael Noll, "Bell Labs," A. Michael Noll, last modified January 14, 2018, http:// noll.uscannenberg.org/BellLabs.html; Noll, "Biography," http://noll .uscannenberg.org/Biography.htm.

*176*  indeed is little more than real fantasy . . . William Dutton, "A Challenge to Virtual Reality by A. Michael Noll," Quello Center, September 26, 2016, http://quello.msu.edu/a-challenge-to-virtual-reality-by-a-michael-noll/.

*177*  the controller vibrated appropriately . . . Martin Watts, "Taking a Look Back at the Nintendo 64 Rumble Pak," NintendoLife, July 11, 2013, http:// www.nintendolife.com/news/2013/07/feature_taking_a_look_back_at _the_nintendo_64_rumble_pak.

*178*  raking its claws across your back . . . Richard Lai, "bHaptics' TactSuit Is VR Haptic Feedback Done Right," Engadget, July 2, 2017, https://www .engadget.com/2017/07/02/bhaptics-tactsuit-vr-haptic-feedback-htc-vive -x-demo-day/.

*178*  who had come up with the Data Glove, which *was* . . . hackertrips, "An Interview with Mitch Altman (Inventor and Virtual Reality Pioneer from the 80's)," YouTube, January 28, 2015, https://youtu.be/5TrRO_j_efg?t= 36m40s.

*178* the relevant finger when you "press" a key . . . Chien-Min Wu, Chih-Wen Hsu, and Shana Smith, "A Virtual Reality Keyboard with Realistic Key Click Haptic Feedback," in *HCI International 2015—Posters' Extended Abstracts* (Cham: Springer, 2015), 232–37, https://doi.org/10.1007/978-3-319-21380-4_41.

*179* what it calls a "whole-body human-computer interface" . . . Jacob A. Rubin and Robert S. Crockett, Whole-body human-computer interface, US Patent 9,652,037, filed December 28, 2015, issued May 16, 2017, http://patft.uspto.gov/netacgi/nph-Parser?Sect1=PTO1&Sect2=HITOFF&d=PALL&p=1&u=%2Fnetahtml%2FPTO%2Fsrchnum.htm&r=1&f=G&l=50&s1=9652037.PN.&OS=PN/9652037&RS=PN/9652037.

*179* The company has announced plans . . . Engadget, "Tactile sensation in VR with HaptX Glove hands-on," YouTube, November 21, 2017, https://www.youtube.com/watch?v=UaWZpRsyZ74.

*179* you can feel its legs skittering on your palm . . . Kent Bye, "AxonVR Is Building a Generalized Haptic Display," Road to VR, June 13, 2017, https://www.roadtovr.com/axonvr-building-generalized-haptic-display/.

*180* could be used to communicate emotion through the air . . . Marianna Obrist et al., "Emotions Mediated Through Mid-Air Haptics," in *Proceedings of the 33rd Annual ACM Conference on Human Factors in Computing Systems* (New York: ACM, 2015), 2053–62, https://doi.org/10.1145/2702123.2702361.

*181* in keeping with the contours of a virtual object . . . Hrvoje Benko et al., "NormalTouch and TextureTouch: High-Fidelity 3D Haptic Shape Rendering on Handheld Virtual Reality Controllers," in *Proceedings of the 29th Annual Symposium on User Interface Software and Technology* (New York: ACM, 2016), 717–28, https://doi.org/10.1145/2984511.2984526.

*182* as one commenter on a YouTube video demonstrating the controllers . . . Smack Thrustcrusher comment on MSPower user, "Microsoft Research: NormalTouch and TextureTouch," YouTube, October 17, 2016, https://www.youtube.com/watch?v=SC5v3u0vmm4.

*182* we are programmatically controlling the user's tactile perception . . . Oliver Bau and Ivan Poupyrev, "REVEL: Tactile Feedback Technology for Augmented Reality," *ACM Transactions on Graphics* 31, no. 4, article no. 89 (2012): 1–11, http://doi.acm.org/10.1145/2185520.2185585.

*187* what it called "teleoperator and virtual environment systems" . . . The Editors, "Welcome to the New Journal," *Presence: Teleoperators and Virtual Environments* 1, no. 1 (1992): iii, https://doi.org/10.1162/pres .1992.1.1.iii.

*187* the first "easter egg" in a video game . . . Warren Robinett and Jannick P. Rolland, "A Computational Model for the Stereoscopic Optics of a Head-Mounted Display," *Presence: Teleoperators and Virtual Environments* 1, no. 1 (1992): 45–62, https://doi.org/10.1162/pres.1992.1.1.45; Jared Petty, "Gaming's First Easter Egg," IGN, March 5, 2015, http://www.ign.com /articles/2015/03/05/gamings-first-easter-egg-adventure-lets-play-with -creator-warren-robinett.

*188* we believe we have made considerable progress . . . Bob G. Witmer and Michael J. Singer, "Measuring Presence in Virtual Environments: A Presence Questionnaire," *Presence: Teleoperators and Virtual Environments* 7, no. 3 (1998): 225–40, https://doi.org/10.1162/105474698565686.

*188* the responses it engendered from other researchers . . . Mel Slater, "Measuring Presence: A Response to the Witmer and Singer Presence Questionnaire," *Presence: Teleoperators and Virtual Environments* 8, no. 5 (1999): 560–72, https://doi.org/10.1162/105474699566477.

*188* couldn't even distinguish between a real experience and a virtual one . . . Martin Usoh et al., "Using Presence Questionnaires in Reality," *Presence: Teleoperators and Virtual Environments* 9, no. 5 (2000): 497–503, https:// doi.org/10.1162/105474600566989.

*189* exploring it from the user's perspective . . . Fabrizio Davide and Richard Walker, "Engineering Presence: An Experimental Strategy," in *Being There: Concepts, Effects and Measurement of User Presence in Synthetic Environments*, ed. G. Riva, F. Davide, and W. A. Ijsselsteijn (Amsterdam: IOS Press, 2003), 41–57, http://citeseerx.ist.psu.edu/viewdoc/summary ?doi=10.1.1.73.6243.

*189* not from the world of VR, but from the world of design . . . Dustin B. Chertoff, Brian F. Goldiez, and Joseph J. LaViola Jr., "Virtual Experience Test: A Virtual Environment Evaluation Questionnaire," in 2010 IEEE Virtual Reality Conference (Los Alamitos, CA: IEEE, 2010), 103–10, https:// doi.org/10.1109/VR.2010.5444804.

*190* there are others as well . . . Kent Bye, "Historical Context of Virtual Reality: What We Can Learn from the Ancients" lecture, Silicon Valley Virtual

Reality Conference and Expo, San Jose, March 29, 2017, YouTube, https://www.youtube.com/watch?v=4rJceG9DweA (32: 40).

190 he started out working on user interfaces at Apple . . . "Emerging Tech in Filmmaking with Director Ryan Staake," The Mill, May 4, 2015, http://www.themill.com/millchannel/403/emerging-tech-in-filmmaking -with-director-ryan-staake.

191 thirty million views on YouTube . . . Young Thug, "Young Thug—Wyclef Jean [Official Video]," YouTube, January 16, 2017, https://www.youtube .com/watch?v=_9L3j-lVLwk.

191 could easily have been the mayor of Uncanny Valley . . . Pomp & Clout, "Young Thug—Wyclef Jean 'Live' Q&A w/Ryan Staake," Vimeo, https://vimeo.com/204094570.


## CHAPTER 9: XXX-CHANGE PROGRAM

197 a large "digital entertainment company" . . . "Welcome to CMP Group," CMP Group, accessed January 21, 2018, http://teamcmp.com/.

200 to more than nine hundred thousand . . . "Virtual Reality Porn," Porn Hub Insights, May 11, 2017, https://www.pornhub.com/insights/virtual -reality.

201 For tens of thousands of years . . . Lawrence Conway, "Aboriginal Erotic Rock Art Proves That—Even 28,000 Years Ago—Men Had ONE Thing on Their Minds," The Daily Mail, June 18, 2012, http://www.dailymail.co.uk /news/article-2161118/Aboriginal-erotic-rock-art-proves-28-000-years -ago-men-ONE-thing-minds.html.

202 male visitors are 160 percent more likely . . . "Virtual Reality Porn."

202 women constituted a mere 26 percent . . . "Pornhub's 2016 Year in Review," Porn Hub Insights, January 4, 2017, https://www.pornhub.com/insights /2016-year-in-review.

202 that majority has only grown . . . "Porn on the Go: Mobile Traffic Take-over," Porn Hub Insights, July 19, 2016, https://www.pornhub.com /insights/mobile-traffic.

203 people spend fewer than 10 minutes per visit . . . "Pornhub's 2016 Year in Review."

203  the rise of MP3s created a precipitous dip in album sales . . . Adrian Covert, "A Decade of iTunes Singles Killed the Music Industry," CNN, April 25, 2013, http://money.cnn.com/2013/04/25/technology/itunes -music-decline/.

203  were filmed in increasingly demeaning ways . . . Katrina Forrester, "Making Sense of Modern Pornography," *The New Yorker,* September 26, 2016, http://www.newyorker.com/magazine/2016/09/26/making-sense -of-modern-pornography.

203  such as slapping, spanking, or gagging . . . Ana J. Bridges et al., "Aggression and Sexual Behavior in Best-Selling Pornography Videos: A Content Analysis Update," *Violence Against Women* 16 (2010): 1065–85, https://doi .org/10.1177/1077801210382866.

207  so that they couldn't see their own arm . . . Matthew Botvinick and Jonathan Cohen, "Rubber Hands 'Feel,' Touch That Eyes See," *Nature* 391 (1998): 756, https://doi.org/10.1038/35784.

207  calling it the "virtual arm illusion" . . . Mel Slater et al., "Towards a Digital Body: The Virtual Arm Illusion," *Frontiers in Human Neuroscience* 2, no. 6 (2008), https://doi.org/10.3389/neuro.09.006.2008.

208  induce the feeling of a full-body transfer . . . Mel Slater et al., "First Person Experience of Body Transfer in Virtual Reality," *PLOS ONE* 5, no. 5 (2010): e10564, https://doi.org/10.1371/journal.pone.0010564.

209  Masters and Johnson's "squeeze" technique . . . Janet S. St. Lawrence and Sudhakar Madakasira, "Evaluation and Treatment of Premature Ejaculation: A Critical Review," *International Journal of Psychiatry in Medicine* 22, no. 1 (1992): 77–97, https://doi.org/10.2190/UWP1-CNHH-L0NK-YQY9.

211  more time to develop the character before sex . . . Maia Szalavitz, "Q&A: The Researchers Who Analyzed All the Porn on the Internet," *TIME,* May 19, 2011, http://healthland.time.com/2011/05/19/mind-reading-the -researchers-who-analyzed-all-the-porn-on-the-internet/.

212  especially among young women . . . Cleo Stiller, "Why Women Love Porn GIFs," Splinter News, July 15, 2015, https://splinternews.com/why-women -love-porn-gifs-1793849213.

217  each furnished with a bed and a webcam . . . Jack Davies, "I Spent a Month Living in a Romanian Sexcam Studio," Vice, December 9, 2013, https://

www.vice.com/en_us/article/mv5e3n/bucharest-webcam-studios
-america-outsourcing-sex-trade.

*218*  if you practice shooting free throws . . . James E. Driskell, Carolyn Copper, and Aidan Moran, "Does mental practice enhance performance?" *Journal of Applied Psychology* 79, no. 4 (1994), 481–92, https://doi.org/10.1037// 0021-9010.79.4.481.


## CHAPTER 10: WHERE WE'RE GOING, WE DON'T NEED HEADSETS

*223*  two Boeing researchers in the early 1990s . . . Thomas P. Caudell and David W. Mizell, "Augmented Reality: An Application of Heads-Up Display Technology to Manual Manufacturing Processes," in *Proceedings of the Twenty-Fifth Hawaii International Conference on System Sciences* 2 (Los Alamitos, CA: 1992): 659–69, https://doi.org/10.1109 /HICSS.1992.183317.

*224*  so your eyes could follow it better . . . Sal Barry, "20 Years Later, a Look at the Foxtrax Puck's Complex Legacy," Hockey News, January 23, 2016, http://www.thehockeynews.com/news/article/20-years-later-a-look-back -at-the-foxtrax-pucks-complex-legacy.

*224*  lab that spun off into its own company . . . Mark Bergen, "Why Did Google Get Rid of the Company Behind Pokémon Go?," Recode, July 12, 2016, https://www.recode.net/2016/7/12/12153722/google-niantic-pokemon -go-spin-out; "Our Team," Niantic, accessed January 21, 2018, https:// www.nianticlabs.com/about.

*225*  bringing *Wizard of Oz* characters into the physical world . . . David Pierce, "Google Joins the Augmented Reality Party with ARCore," *WIRED*, August 29, 2017, https://www.wired.com/story/google-joins-the -augmented-reality-party-with-arcore/.

*226*  and throwing an eighty-inch TV on the wall . . . CNET, "Andre Iguodala Has Tried Magic Leap. What's It Like?," YouTube, April 13, 2017, https:// www.youtube.com/watch?v=YPjc7qAXwy8.

*226*  glowing appendage to touch my fingertip . . . Kevin Kelly, "Hyper Vision," *WIRED* (May 2016): 74–87, 112, https://www.wired.com/2016/04/magic -leap-vr/.

*229*  an AR headset to be released in 2020 . . . Mark Gurman, "Apple Is Ramping Up Work on AR Headset to Succeed iPhone," Bloomberg News, November 8, 2017, https://www.bloomberg.com/news/articles/2017-11-08/apple-is -said-to-ramp-up-work-on-augmented-reality-headset.

*229*  then render them inside your headset . . . Michael Abrash, "Inventing the Future," Oculus, October 11, 2017, https://www.oculus.com/blog /inventing-the-future/.

# INDEX

# ABOUT THE AUTHOR

Peter Rubin leads *WIRED*'s editorial efforts on digital platforms and oversees culture coverage in the magazine and online. Prior to arriving at *WIRED* in 2011, he was a feature writer and editor for more than a decade, penning cover stories for *Elle, Details, GQ, Good,* and other magazines. Rubin holds a master's degree from Columbia University and graduated from Williams College. He lives in Oakland, California, with his wife.